Global Perspectives on Health Geography

Series Editor
Valorie Crooks, Department of Geography
Simon Fraser University
Burnaby, BC, Canada

Global Perspectives on Health Geography showcases cutting-edge health geography research that addresses pressing, contemporary aspects of the health-place interface. The bi-directional influence between health and place has been acknowledged for centuries, and understanding traditional and contemporary aspects of this connection is at the core of the discipline of health geography. Health geographers, for example, have: shown the complex ways in which places influence and directly impact our health; documented how and why we seek specific spaces to improve our wellbeing; and revealed how policies and practices across multiple scales affect health care delivery and receipt.

The series publishes a comprehensive portfolio of monographs and edited volumes that document the latest research in this important discipline. Proposals are accepted across a broad and ever-developing swath of topics as diverse as the discipline of health geography itself, including transnational health mobilities, experiential accounts of health and wellbeing, global-local health policies and practices, mHealth, environmental health (in)equity, theoretical approaches, and emerging spatial technologies as they relate to health and health services. Volumes in this series draw forth new methods, ways of thinking, and approaches to examining spatial and place-based aspects of health and health care across scales. They also weave together connections between health geography and other health and social science disciplines, and in doing so highlight the importance of spatial thinking.

Dr. Valorie Crooks (Simon Fraser University, crooks@sfu.ca) is the Series Editor of Global Perspectives on Health Geography. An author/editor questionnaire and book proposal form can be obtained from Publishing Editor Zachary Romano (zachary.romano@springer.com).

More information about this series at https://link.springer.com/bookseries/15801

Ryan J. Petteway

Representation, Re-Presentation, and Resistance

Participatory Geographies of Place, Health, and Embodiment

Springer

Ryan J. Petteway
OHSU-PSU School of Public Health
Portland State University
Portland, OR, USA

ISSN 2522-8005 ISSN 2522-8013 (electronic)
Global Perspectives on Health Geography
ISBN 978-3-031-06140-0 ISBN 978-3-031-06141-7 (eBook)
https://doi.org/10.1007/978-3-031-06141-7

This Springer imprint is published by the registered company Springer Nature Switzerland AG
The registered company address is: Gewerbestrasse 11, 6330 Cham, Switzerland

to 805
Market Square

where it started at

no fitted
just some Payless

and a Starter cap

white sugar
Flavor-Aid

and the water tap

hallways
and night skies

to marvel at

black stars
with black scars

as artifacts

the black top
is a black box

those are the facts

Preface

I remember sitting on the backs of project benches, sipping my orange drink while listening to Mobb Deep ask "where ya from?" and discuss dropping a "gem on 'em." I was curious. Deep thoughts ensued … Huh. A gem, you say? What does this gem entail, exactly? Is it expensive? Can it be sold at the pawnshop? Will it multiply if I get it wet? Is it a metaphysical, theoretical gem—perhaps something RZA would wear to a cocktail reception at the World Congress of Philosophy? Are we dealing with conceptual jewelry here? Also, what and/or where exactly is this "where" they seem to be so fascinated with? Is it a geographic where, or a cartographic where? Rectangle or square? Or perhaps a sociologic where, where people *know* where? And what does it mean to be "from" this "where"? So many questions, so little orange drink to answer them all.[1]

As a public health researcher with a focus on notions of "place," my work is informed by a mix of disciplines, including social epidemiology, geography, and sociology. Importantly, it's also informed by my own lived experiences of place, as well as my practice-based work as a social epidemiologist at a local health department (aspects of which are discussed in Chap. 2). And it's the combination of these that ultimately shaped the foundations of my work as a place-health research scholar. My work at the health department, which involved extensive analysis and GIS mapping of health and social determinants data at the "neighborhood" level (i.e., census tracts), helped me get a deeper understanding of place-based and spatial aspects of community health and the importance of place-health metrics. More importantly, it made me realize how disrespectful, disempowering, settler-colonial, and detached much of such work is—including much of what I had done. Essentially, I got (under)paid as a credentialed (social) epidemiologist to sit by myself in a windowless, air-conditioned office and analyze secondary data on mortality, morbidity, income, education, housing, energy, greenspace, food environment, tobacco and alcohol retail/licensing, non-fatal shootings, rodent and pest complaints, lead exposure, and so on and so forth. I got (under)paid to produce maps based on these analyses. Maps that supposedly represented something true about life, death, and health within the city's neighborhoods. Choropleth maps that literally threw shade

---

[1] Excerpt from Petteway, R. (2018). The Real Limits of Census Tracts, and Other Boundaries. *Shelterforce*.

on peoples' neighborhoodss—many of which I had never actually stepped foot in. I got (under)paid to tell data stories about other peoples' "place."

> And I thought I was killing it … until I remembered that prior to being a social epidemiologist, I listened to Mobb Deep and drank orange drink while conducting observational studies of project life.[2]

Somehow, long ago, far far away… in the world of public health and epidemiological surveillance, it became standard procedure for credentialed technocrats to have unilateral control over both the data about peoples' places and the narratives produced about those places using the data. No questions asked. Government data analysts extracting and compiling peoples' data and forging spatial narratives of their communities that serendipitously appear in time to buttress whichever initiative the current regime decided was important. What I was a part of was disrespectful at best. I even suggested to our health commissioner at the time that we supplement "our" data with community-generated data and develop neighborhood data, organizing, and advocacy hubs co-led by residents—so that we would have more relevant, real-time, and actionable data and communities could have capacity to identify, analyze, and map their own concerns: so that they would have agency and control over their own neighborhood data story—the power to co-create *their* place-health narrative. It was implied that I would be fired if I pursued such an endeavor—because that's "too political."

And that, perhaps above all else, was the crux of the matter: this agency was operating under the illusion that its data practices were somehow not inherently and inescapably political from day one. Which data are collected? By whom? Who gets paid to collect and analyze the data? Who has access to the data? Who does not? Who owns the data? How much does it cost? What counts as data? Whose data counts? All of these questions are not only political questions, but in the context of place and health they are matters of policy and politics themselves. The power relations that govern the presence/absence, available/unavailability, and accessibility/inaccessibility of data about place are arguably the most fundamentally political matter in a city. For a health department to feign ignorance of this—or, to willfully align itself with the lie of apolitical neutrality in data/research practice—runs counter to every core principle of health equity and justice. How can a health department say it is advancing health equity in peoples' neighborhoods if they view producing data narratives and maps about those neighborhoods as apolitical and power-detached acts of perfunctory surveillance duty? Without even—at minimum—questioning the functions, impacts, and choices of what is being surveilled and why? Yes, I was relatively new to the field then, but even I appreciated (even if still complicit) that data narratives and maps, as visual codifications of spatialized power, "are neither neutral nor unproblematic with respect to representation, positionality, and partiality of knowledge" (Harris & Hazen, 2005, p. 101). Rather, as articulated by Willow (2013), "because they depict a subjective and specific view of the world rather than any objective external reality, maps can also be used to challenge

---

[2] Ibid. 1.

dominant ways of conceiving the landscape and the socio-political interests they represent" (p. 872). So, why wouldn't—and shouldn't—it have been standard practice to do what I suggested?

I took this question—along with many others—with me when I resigned and pursued doctorate training. Years later—now as a doubly credentialed place-health scholar at the intersections of social epidemiology, geography, and sociology—I still have no answers. Just more questions. But I've learned that much existing work at these intersections, especially within public health (and most especially epidemiology), has tended to favor highly quantitative and "outsider"-driven approaches to understanding place-health contexts and relationships—for example, non-participatory survey-based research conducted by outside researchers without collaboration or meaningful engagement with/within the communities they're researching. Community residents—as research participants/"n's"—are viewed and valued primarily *as* data points, and not as political constituents and social agents that could help explain or intervene *on* the data and its determinants. This orientation, for me, is very much rooted in the dominant paradigm of public health research—one of positivist, reductionist, White supremacist, and settler-colonial logics of extraction and expropriation which habitually fails to engage matters of power, epistemic justice, procedural justice, and distributive justice as germane to knowledge production. Procedural and methodological norms under traditional place-health research approaches preclude expressions of/actively mask participants' agency, discounting/devaluing their knowledge/expertise while simultaneously dispossessing them of their stories/experiences ("data"). In this regard, relationships between researchers and participants/residents represent not only the re-inscription of social hierarchy but also the reification of place-health research as (re)colonization. A consequent analytic concern is the production of spatial narratives that mischaracterize/misrepresent important community spaces and experiences, affording only partial and decontextualized—ahistoric, apolitical, de-humanized—renditions of place-health geographies.

In contrast, one of the defining elements of my work has been an emphasis on inclusion (e.g., epistemic and procedural justice) in the place-health knowledge production process—viewing place-health research as a core component of placemaking. My work is collaborative and employs exclusively participatory methods and processes to elucidate peoples' lived geographies of place and health. In doing so, I approach efforts to unpack place-health relationships and outline geographies of health therein as an iterative dialogue between residents and researchers. Thus, for me, the growing prominence of community-based participatory research (CBPR) and increasing utility/uptake of information and communication technologies (ICTs) afford opportunities to democratize—and decolonize—place-health research and practice.

In this spirit, for this book, I draw from roughly 10 years of participatory work to examine core themes of *(mis)representation*, *re-presentation,* and *resistance* within place-health research and practice. In doing so, I recount, report, and reflect on a selection of place-focused health research projects that I have developed over the years, including practice-oriented health department work examining local food environments among youth in public housing (Chap. 2), and intergenerational participatory research exploring "geographies of embodiment" with residents of public housing (Chap. 5). Each chapter highlights key concerns/considerations in regard to place (mis)representation, and highlights the role/value of inclusive participatory methods and processes (e.g., CBPR, ICTs) in re-presenting—and decolonizing—spatial narratives of health. Chapters also touch on critical notions of place-health research as a potential mode of *resistance* in the production of spatial knowledge and narratives of "place." The chapters discuss a mix of practice- and research-based projects with implications and applications for practitioners (e.g., local health department epidemiologists) and academics, introducing readers to an array of new and mixed-methods (Chaps. 3, 4, and 5). In doing so, this book emphasizes inclusive, participatory methods and processes rooted in decolonizing, CBPR, feminist, and Black feminist principles, and introduces readers to a new place-health research framework (Chap. 3). Overall, this book offers an integrated call and framing for a critical examination of how "place" and health geographies—and narratives/stories therein—are constructed, and perhaps might be de/re-constructed through inclusive and equitable research practices—capable, perhaps, of capturing and honoring the power of an orange-drink-sipping, Mobb-Deep-soundtracked moment of place beauty buried beneath a stranger's choropleth of naturally broken death and pretty decay.

Portland, OR, USA   Ryan J. Petteway

References

Harris, L. M., & Hazen, H. D. (2005). Power of maps: (Counter) Mapping for conservation. *ACME: An International Journal for Critical Geographies, 4*(1), 99–130.
Willow, A. J. (2013). Doing sovereignty in native North America: Anishinaabe counter-mapping and the struggle for land-based self-determination. *Human Ecology, 41*(6), 871–884. https://doi.org/10.1007/s10745-013-9593-9

Contents

List of Figures

List of Tables

About the Author

Ryan J. Petteway is a social epidemiologist and assistant professor in the OHSU-PSU School of Public Health in Portland, OR. His applied research integrates social epidemiology, critical theory, decolonizing methods, and community-based participatory research (CBPR) to examine notions of place, embodiment, and placemaking in community health and development, making use of information and communication technologies (ICTs) to democratize research and practice processes.

More broadly, his scholarship engages: (1) notions of epistemic, procedural, and distributive justice within public health knowledge production processes—including considerations of power and (mis)representation in data collection, analysis, and use; (2) applications of critical theory to examine dominant discourse/narrative frames of "health equity," for example, considerations of power and epistemic equity; and (3) pervading ethical frames of public health law and police powers.

Dr. Petteway is also an award-winning poet, including a National Poetry Month Prize, a Paper of the Year Award, and a Pushcart Prize nomination. His poems have appeared in both peer-reviewed academic journals and traditional poetry presses.

He is an alumnus of the University of Virginia, the University of Michigan, and the University of California, Berkeley.

Chapter 1
Refining and (Re)Defining "Place" in Health Research: Interrogating Spatial Knowledges + (Mis)Representations

Introduction

Understanding how "place," e.g., neighborhoods/community environments, affects health is a prominent concern within public health (Arcaya et al., 2016; Diez Roux & Mair, 2010; Santos et al., 2007). While research has grown rapidly over the last two decades, many conceptual, methodological, and procedural challenges remain. Conceptually, "place" more often than not is defined as an administrative location of residence, e.g., the census tract where one lives, or as some other "territorial neighborhood" (Arcaya et al., 2016; Chaix et al., 2009). Additionally, most place-health research views "place" as singular in nature, thus the census tract ("neighborhood") of residence is the *only* place examined in most studies (Chaix et al., 2009; Diez Roux & Mair, 2010). However, a growing body of literature has made clear that efforts to unpack relationships between place and health must be able to account for people's spatially-specific place experiences (Chaix et al., 2009; Cummins et al., 2007; Cutchin et al., 2011; Macintyre et al., 2002; Matthews, 2011; Rainham et al., 2010). Moreover, place-health work to date has done little to account for/unpack how place is actively made, unmade, and remade (e.g., via placemaking mechanisms like redlining, urban renewal, and gentrification), which limits the extent to which findings can inform social and policy action.

As a public health researcher with a focus on notions of "place," my work is informed by a mix of disciplines, including social epidemiology, geography, and sociology. Existing work at these intersections, especially within public health, has tended to favor highly quantitative and "outsider"-driven approaches to understanding place-health contexts and relationships—e.g., non-participatory survey-based research conducted by outside researchers without collaboration or meaningful engagement with/within the communities they're researching. Community residents—and/as research participants—are viewed and valued primarily *as* data points, not as political constituents with agency and networked

R. J. Petteway, *Representation, Re-Presentation, and Resistance*, Global Perspectives on Health Geography, https://doi.org/10.1007/978-3-031-06141-7_1

social power that could help explain or intervene *on* the data and its determinants. In this way, procedural and methodological norms under traditional approaches preclude expressions of/actively mask residents'/participants' agency, discounting/devaluing their knowledge/expertise while simultaneously dispossessing them of their stories/experiences ("data"). In this regard, relationships between researchers and residents represent not only the re-inscription of social hierarchy, but the reification of place-health research *as* (re)colonization. A consequence and analytic concern is the production of spatial narratives that not only misspecify place-health effects but mischaracterize/misrepresent important community spaces, affording only partial and decontextualized—ahistoric, apolitical, disembodied—renditions of place-health geographies. Accordingly, a core and animating question for this book is, "How can we reimagine a more dynamic, epistemically just, and actionable place-health research—one which can counter/enhance existing place-health (mis)representations, serve as a process for re-presentation, and act as a site/mode of critical resistance?"

In this regard, I suggest that there are perhaps two broad and interconnected areas that impede advancement of sounder place-health science and decolonized, actionable place-health knowledge. First, there is a great need for conceptual, procedural, and methodological improvement to better account for the dynamic multi-nodal nature of "place" and its contingent spatial, temporal, and social internodal connections (and systemic, structural divisions) and to elucidate potential intergenerational and life-stage differences in place experiences/perceptions. Second, there has been an absence of critical discourse that expressly engages considerations of power and (mis)representation within place-health research and knowledge production, leaving unquestioned and unchallenged default research orientations and practices that, by and large, render place-health geographies void of social, political, and historic context. In other words, there is a great need for community-inclusive, decolonized work that explicitly engages the sociopolitical mechanisms that make, unmake, and remake place over time, i.e., social and material placemaking processes (including place-health knowledge production), thus shaping spatiotemporal patterns and sociospatial arrangements of place exposures and opportunities.

In the following sections, I summarize some core conceptual, procedural, and methodological challenges and opportunities within place-health research and briefly discuss the potential value of intergenerational and participatory approaches and methods to advance efforts to democratize—and decolonize—spatial narratives and representations of place-health geographies. I then outline the remaining chapters, each of which, in drawing from participatory and intergenerational research projects, speak to themes of (mis)representation, re-presentation, and resistance within place-health knowledge production.

(Mis)Representation

What Is This Place? Conceptual Challenges and Opportunities in Place-Health Research

Many conceptual and methodological challenges remain in place-health research (Bernard et al., 2007; Cummins et al., 2007; Cutchin, 2007; Diez Roux, 2004; Diez Roux & Mair, 2010; Frumkin, 2006; Kwan, 2009; Macintyre et al., 2002; Matthews, 2008, 2011; Mujahid et al., 2007; Rainham et al., 2010; Spielman & Yoo, 2009). Of particular prominence are matters related to conceptualizing and defining "place" and accounting for changes in place over time. For example, "place" is almost exclusively defined as an administrative location of residence, e.g., census tract where one lives, or as some other "territorial neighborhood" (Arcaya et al., 2016; Chaix et al., 2009; Diez Roux & Mair, 2010; Riva et al., 2008). Additionally, most place-health research views "place" as singular in nature, thus the census tract ("neighborhood") of residence is the *only* place examined in most studies. Moreover, the majority of place-health work has been cross-sectional, meaning that our under-standing of place and health is derived mostly from examinations of one singular place at only one particular point in time. In short, "place" in health research has largely been arbitrary, singular, static, and, perhaps most importantly, operationally invisible and meaningless to those residing in it—such a place does not exist in their lived reality. Many researchers have, of course, questioned the legitimacy and utility of such a conception of and approach to studying place, and some have suggested more appropriate approaches. Among important contributions that I draw from in this book are notions of "opportunity structures" and needs-driven place configura-tions (Macintyre et al., 2002), "relational" place (Cummins et al., 2007), "spatial polygamy" (Matthews, 2011), and especially activity space (Browning & Soller, 2014; Chaix et al., 2012; Jones & Pebley, 2014; Kwan, 2009; Perchoux et al., 2013).

As noted by Matthews (2008, p. 259), there is "abundant evidence that people jump spatial scales and move across multiple, non-nested hierarchies in their daily activities." That is, people are not bounded by the artificial lines we often use to define place in our studies, thus our measures of place-effects are quite haphazard—presuming that the only place-based exposures of importance for health occur "24/7/365" in one location (Kwan, 2009; Matthews, 2008). Macintyre et al. (2002) suggest an approach to place that is rooted in an understanding of human needs and how they are met. A particular location, a "neighborhood" for example, will only provide some of the requisite "opportunity structures" needed to support and sustain a healthy life; thus, people will inevitably have to navigate to and through multiple places. And it's critical to understanding that such opportunity structures are socially and politically constructed features of communities that, in their presence or absence, either directly or indirectly constrain possibilities for health. Accordingly, "this means operationalizing measures, appropriate for the particular society and historical period, of the ways in which these needs are met in particular places" (Macintyre et al., 2002, p. 133). Implicit in this approach is that each person will have a unique set of needs that they will need to meet in a particular way; thus, any notion of "place" and place-effects must be able to account for similarities and differences between peoples' needs-driven configuration of places—and these places are not in one singular area. Additionally, the authors encourage recognition of the dynamic and changing reality of places and peoples' interaction with features of place over time, and the need for theorizing around time in place-health research.

Cummins et al. (2007) put forth a "relational" approach to understanding place. This approach emphasizes the need to view "place" as more than just an unchanging geographic location—rather, it is a dynamic social and cultural production inextricably linked to political and economic processes that unfold on various spatial scales (e.g., county, city, neighborhood). This means, we need to account for how people perceive and relate to their daily social and physical environments and identify the political and economic forces that shape those environments. From this perspective, places are best conceptualized as nodes within networks that are connected and/or separated by "socio-relational" distance. Additionally, these nodes and their bounds are seen as fluid and dynamic, changing over time. Moreover, people are no longer viewed as agency-less entities within a fixed area but as actors with variant mobility patterns over time, e.g., a day, a week, their lifecourse.

Similarly, work by Matthews et al. (2005) suggests understanding "place" based on the multiple spatial locations people interact with during their daily activities— the heterogeneity of peoples' daily places a testament to the reality of their "spatial polygamy" (Matthews, 2011). From this perspective, "place" is no longer a singular location, but rather a particular configuration of nodes that constitutes a spatiotemporal network of a person's lived reality of multiple locations (Matthews, 2011; Matthews et al., 2005; Matthews & Yang, 2013). People have meaningful relationships with multiple nested and non-nested places simultaneously, and these relationships tend to be both an element of, and structured by, present and historic geographic and social contexts. Thus, to appropriately conceptualize and measure place, we must account for person-centered spatial configurations of multiple concurrent place attachments, as well as the forces which have historically shaped and presently drive/maintain such configurations. Instead of arbitrarily defining place as a singular static and largely imaginary set of bounds, "place" becomes a dynamic reflection of peoples' real spatial experiences—experiences that are shaped by and possess historical, social, and political meanings (Albright et al., 2011; Cummins et al., 2007; Kemp, 2011; Macintyre et al., 2002; Matthews, 2008).

The concept of "spatial polygamy," much like relational place, thus aptly captures the multi-nodal nature of a person's lived place—moving from, to, and through place to place throughout a day or week, for example (Matthews, 2011). The location of residence is but one node, and each individual's configuration of non-residential nodes will be different. Accordingly, the spatial polygamy approach extends the idea of "ego-centered neighborhoods"—each person's "place" ("neighborhood") becomes the aggregate of their person-centered nodes and inter-nodal connections (Chaix et al., 2009; Matthews, 2011). Because it does not artificially bound peoples' experience of place contexts, it enables accounting for and responding to health-related exposures beyond simply the "neighborhood," thus avoiding the "local trap" that pervades most place-health research (Cummins, 2007).

Relatedly, attention has also been drawn towards the relevance of concepts from time geography, specifically, person-centered time-space and activity space (Browning & Soller, 2014; Chaix et al., 2012; Hägerstrand, 1970; Jones & Pebley, 2014; Kwan, 2009; Perchoux et al., 2013; Rainham et al., 2010). Here, "place" is less about a specific fixed location (e.g., neighborhood) and more about a specific person's actual daily "action space" (Kwan, 2009). Accordingly, notions of neighborhood and how to define it become largely irrelevant—a person's "place" is defined by where they go and how much time they spend en route and once they get there—an understanding complementary to that suggested by Macintyre et al. (2002) regarding human needs and "effective neighborhoods," and "relational" place as described by Cummins et al. (2007).

Within an activity space approach, we can begin to appreciate the importance of elucidating, for example, considerations of "fixed" versus "flexible" places (Perchoux et al., 2013), and the importance of various space-time constraints— "capability," "coupling," and "authority" constraints (Hägerstrand, 1970)—on people's movements and consequent place-heath geographies. No two peoples' place-time experiences and exposures will be exactly alike (presumably), owing to a combination of oft-ignored relational sociospatial and temporal contexts. Thus, "place" effects on health are best viewed as the product of the space- and time-specific exposures people encounter in the course of their daily lives—multiple places for varying amounts of time, with those places and timings structured beyond simple choice and volition. As posited by Rainham et al. (2010, p.169), "place-based health research would benefit from both a greater knowledge of the patterns of movements of people, and insight into the heterogeneity of context associated with these movements within the population of interest." Within this view, place becomes person-centered and time-bound in character—as opposed to strictly location-centered and timeless—and must be situated within appropriate historic and sociospatial contexts.

Re-Presentation

Procedural and Methodological Challenges and Opportunities: A Case for Intergenerational and Participatory Place-Health Research

In addition to core conceptual challenges, there are two central procedural limitations in the majority of place-health research to date that represent opportunities to enhance the field, especially in regard to work having the potential to inform action within the contexts where research unfolds. First, a growing body of work suggests the need to account for multigenerational place-effects (Galster & Sharkey, 2017).

For example, neighborhood deprivation has been shown to be associated with child cognitive development across generations (Sharkey & Elwert, 2011). This work suggests that research must consider both "direct and indirect pathways by which neighborhood exposures in both the parent and child generations may influence children's trajectories" (p. 2). While there is a general consensus that such exposures indeed exert an influence on health and developmental outcomes, the potential mechanisms of multigenerational effects remain unclear.

Additionally, elaboration will require an improved understanding and enumeration not only of stable and varying exposures *across* generations, but also how perceptions of these exposures might vary *between* generations. Unfortunately, there is a paucity of place-health work that incorporates both youth and adult perspectives or accounts for changes in "place" over time. As public health practice and research continue to evolve and become more sensitive to the need for a lifecourse perspective on health (Ben-Shlomo & Kuh, 2002; Hertzman et al., 2001; Hertzman & Power, 2003; Lynch & Smith, 2005), and the role of place contexts and place embodiment (Curtis et al., 2004; Gustafsson et al., 2014; Merlo, 2011; Nazmi et al., 2010; Petteway et al., 2019a), it will become increasingly important to include opportunities for intergenerational perspectives and participation in the work. Place-based experiences and perceptions thereof are inextricably linked to age and life-stage, an especially significant consideration given that subjective measures of place (not just objective ones) matter for health (Barrington et al., 2014; Lin & Moudon, 2010; Pruitt et al., 2012; Schulz et al., 2013; van Deurzen et al., 2016; Weden et al., 2008; Wen et al., 2006). Adults and youth have fundamentally different place experiences, encountering different physical and social environments throughout their day, for example, and those environments change over time. And a growing body of work has done well to identify the significance of age/life-stage considerations in understanding place-health relationships (Fang et al., 2016; Hand et al., 2018; John & Gunter, 2016; Milton et al., 2015; Tong et al., 2016; Villanueva et al., 2012; York Cornwell & Cagney, 2017). The intentional inclusion of more real-time youth perspectives in current place-health work would enhance retrospective and real-time work with adults and allow for a more rigorous and nuanced examination of place and health across generations and across the lifecourse. Moreover, an intergenerational approach can facilitate examination of how perceptions vary between youth and adults, and how these perceptions change over time—thus improving our ability to identify and appropriately characterize exposures, and correctly specify multigenerational mechanisms.

Second, most public health work examining place and health to date has not effectively incorporated participatory approaches/methods. This practice gap presents as a missed opportunity to critically engage residents in the process of defining, understanding, and changing place and its local/regional policy/political determinants, considerations for "scaling up" withstanding (Bambra et al., 2019). This chasm has been highlighted for redress within social epidemiology, place-health, and participatory public health research literature (Lantz et al., 2006; Leung, 2004; Petteway et al., 2019b; Wallerstein et al., 2011). Incorporating participatory methods within place and health research can help bridge this gap. Employing an approach that is rooted in community-based participatory research (CBPR) and makes use of participatory methods can help ensure local knowledge and expertise are prioritized within the research process, and facilitate power-sharing and critical engagement among local communities, research participants, and outside researchers (Israel et al., 1998, 2010; Minkler, 2000, 2010; Wallerstein & Duran, 2010). Thus, the research findings reflect nuances and perspectives of peoples' lived realities that otherwise are often missed using non-participatory methods and a non-participatory approach. Such an integrated approach can allow for a more organic, grounded, and locally relevant exploration of place that can better inform place-health theory and metric development, as well as the development of place-based public health strategies. A community-engaged approach that utilizes participatory process and methods can improve not only research on place and health (e.g., research questions, data collection, analysis, and dissemination), but also local research translation and action based on the work completed (Balazs & Morello-Frosch, 2013; Minkler, 2010; Morello-Frosch Jr et al., 2005; Wallerstein & Duran, 2010).

Methodologically and procedurally, most place-health research to date has been non-participatory survey-based quantitative work, with only limited use of community-inclusive processes or methods (Cannuscio et al., 2009; Carpiano, 2009; Dennis et al., 2009; Richardson & Nuru-Jeter, 2012; Schultz et al., 2005). However, place-health research is well-suited to incorporate and benefit from community-inclusive approaches, particularly CBPR. Such integration is an opportunity to leverage the *practical and procedural* translational advantages of much place-based research (e.g., space-bound, locality- and jurisdictionally-specific), while simultaneously capitalizing on the *scientific and political* translational advantages of harnessing place-based knowledge, insight, and expertise of the people whose lives unfold within the "place" being studied. And critically, such approaches can facilitate critical application of decolonized orientations to place-health research, moving beyond transactional data extraction (and abstraction) processes that commodify, decontextualize, and disembody data for the production of spatial narratives that de-center lived and embodied local knowledges. While notions of decolonizing have been articulated within broader public health discourses (Chandanabhumma & Narasimhan, 2020; Darroch & Giles, 2014; McGibbon et al., 2014; Mundel & Chapman, 2010), there appears to be an absence of place-health research that expressly centers decolonizing theoretical orientations and methods, or considerations for decolonization and settler-colonialism within place-health knowledge production. For public health at least, the prominence and traction of CBPR afford an opportunity, and architecture, to more deliberately bring these considerations into the fold of place-health work (something I return to in Chap. 6).

Resistance

Participatory Research Geographies of Place, Health, and Embodiment: Organization of This Book

In the following chapters, I examine how participatory, community-inclusive approaches to place-health research can improve our work on conceptual, procedural, and methodological fronts, and in doing so, help to address concerns regarding misspecification, to counter misrepresentation, and to resist epistemic erasure. In each chapter, I engage the above areas of consideration by presenting and discussing a selection of participatory place-health research and practice projects that I've completed over the years. Each chapter is organized around the three broad themes of representation, re-presentation, and resistance, drawing from the above body of work, as well as social epidemiology, critical, Black feminist, and decolonizing theory, to nuance and deepen our understanding of place and health.

In Chap. 2, I engage questions/tensions around spatial (mis)representations of place-health experiences related to food environments, specifically from an applied/practice-oriented perspective. I draw from practice-based photovoice work as a social epidemiologist at a local health department (LHD) to examine notions of spatial (mis)representation and the manner in which data and maps—particularly when created by "outsider" administrators—can function to produce spatial narratives that do not adequately or accurately reflect peoples' lived experiences of place. In doing so, I suggest a real need for assessment practices that center community knowledges in (re)defining place-health geographies related to food—knowledges that, unlike administrative representations, bring to the forefront the historic and present sociopolitical contexts that shape inequitable food environments.

In Chap. 3, I draw from critical theory, place-health, geography, critical race, and sociology literatures to introduce the "placescape"—a framework developed to explicitly engage the sociopolitical mechanisms/forces that make, unmake, and remake "place" over time. I outline six core placescape tenets and present a "pilot test" of the placescape via an intergenerational participatory research project examining place, embodiment, and health among public housing residents—the *People's Social Epi Project*, or PSEP. In doing so, I highlight the potential value of intergenerational approaches for deepening/mobilizing resident participation in "placemaking" processes, and the role of participatory research as resistance/for representation within contested urban spaces and place-based health/housing strategies and assessments thereof.

In Chap. 4, also drawing from PSEP work, I describe the application of a multi-method participatory activity space approach to mapping and examining place and health. Informed by the growing body of work calling for relational and spatially dynamic approaches within place-health research, I summarize a fully participatory approach to activity space mapping and discuss implications for community-inclusive methods in improving conceptual and empirical understanding of how place impacts health. In doing so, this chapter speaks to concerns within public health and health geography regarding misspecification of place-health effects and misrepresentation of place-health experiences, with a specific note of the potential value of intergenerational approaches and the use of information and communication technologies (ICTs) to facilitate and center/amplify community place-health knowledge.

In Chap. 5, again drawing from PSEP, I introduce X-Ray Mapping as a new participatory mapping method for elucidating "geographies of embodiment." With a specific focus on LHD contexts—within which social epidemiology, place-health, and GIS capacity are often absent and resident voice is seldom included—I introduce and summarize an application of X-Ray Mapping as resistance/counternarrative to traditional place-health research and LHD assessment approaches. In doing so, I again highlight the potential value of ICTs in enabling/deepening resident participation in assessment efforts and discuss the role of participatory place-health research and practice as resistance/for representation, as well as a potential process to facilitate civic engagement and amplify community voice around matters of equitable and inclusive LHD practice.

In Chap. 6, the concluding chapter, I begin to outline a vision for how to move towards a decolonized, epistemically just approach to place-health research. In doing so, I emphasize the necessity of centering considerations of power in/and the role of *placemaking*—defined as the social, political, and economic policies, processes, and practices—including knowledge production—that actively make, unmake, and remake place overtime, shaping the spatial and relational contexts of health risks and opportunities and narratives thereof. Expanding upon the themes of representation, re-presentation, and resistance touched upon throughout, and extending conceptual considerations articulated via the placescape in Chap. 3, I draw from Black geographies, decolonizing, critical theory, critical race, and Black feminist literatures to outline a conceptually rich scaffold upon/from which place-health scholars can build/draw—one rooted in a praxis of re-presentation, resistance, and reclamation.

References

Albright, K., Chung, G., De Marco, A., & Yoo, J. (2011). Moving beyond geography: Health practices and outcomes across time and place. In L. M. Burton, S. A. Matthews, M. Leung, S. P. Kemp, & D. T. Takeuchi (Eds.), *Communities, neighborhoods, and health* (pp. 127–143). Springer. https://doi.org/10.1007/978-1-4419-7482-2_8

Arcaya, M. C., Tucker-Seeley, R. D., Kim, R., Schnake-Mahl, A., So, M., & Subramanian, S. V. (2016). Research on neighborhood effects on health in the United States: A systematic review of study characteristics. *Social Science & Medicine, 168*, 16–29. https://doi.org/10.1016/j.socscimed.2016.08.047

Balazs, C. L., & Morello-Frosch, R. (2013). The Three Rs: How community-based participatory research strengthens the rigor, relevance, and reach of science. *Environmental Justice, 6*(1), 9–16. https://doi.org/10.1089/env.2012.0017

Bambra, C., Smith, K. E., & Pearce, J. (2019). Scaling up: The politics of health and place. *Social Science & Medicine, 232*, 36–42. https://doi.org/10.1016/j.socscimed.2019.04.036

Barrington, W. E., Stafford, M., Hamer, M., Beresford, S. A. A., Koepsell, T., & Steptoe, A. (2014). Neighborhood socioeconomic deprivation, perceived neighborhood factors, and cortisol responses to induced stress among healthy adults. *Health & Place, 27*, 120–126. https://doi.org/10.1016/j.healthplace.2014.02.001

Ben-Shlomo, Y., & Kuh, D. (2002). A life course approach to chronic disease epidemiology: Conceptual models, empirical challenges and interdisciplinary perspectives. *International Journal of Epidemiology, 31*, 285–293.

Bernard, P., Charafeddine, R., Frohlich, K. L., Daniel, M., Kestens, Y., & Potvin, L. (2007). Health inequalities and place: A theoretical conception of neighbourhood. *Social Science & Medicine, 65*(9), 1839–1852. https://doi.org/10.1016/j.socscimed.2007.05.037

Browning, C. R., & Soller, B. (2014). Moving beyond neighborhood: Activity spaces and ecological networks as contexts for youth development. *Cityscape (Washington, DC), 16*(1), 165.

Cannuscio, C. C., Weiss, E. E., Fruchtman, H., Schroeder, J., Weiner, J., & Asch, D. A. (2009). Visual epidemiology: Photographs as tools for probing street-level etiologies. *Social Science & Medicine, 69*(4), 553–564. https://doi.org/10.1016/j.socscimed.2009.06.013

Carpiano, R. M. (2009). Come take a walk with me: The "Go-Along" interview as a novel method for studying the implications of place for health and well-being. *Health & Place, 15*(1), 263–272. https://doi.org/10.1016/j.healthplace.2008.05.003

Chaix, B., Merlo, J., Evans, D., Leal, C., & Havard, S. (2009). Neighbourhoods in eco-epidemiologic research: Delimiting personal exposure areas. A response to Riva, Gauvin, Apparicio and Brodeur. *Social Science & Medicine, 69*(9), 1306–1310. https://doi.org/10.1016/j.socscimed.2009.07.018

Chaix, B., Kestens, Y., Perchoux, C., Karusisi, N., Merlo, J., & Labadi, K. (2012). An interactive mapping tool to assess individual mobility patterns in neighborhood studies. *American Journal of Preventive Medicine, 43*(4), 440–450. https://doi.org/10.1016/j.amepre.2012.06.026

Chandanabhumma, P. P., & Narasimhan, S. (2020). Towards health equity and social justice: An applied framework of decolonization in health promotion. *Health Promotion International, 35*(4), 831–840. https://doi.org/10.1093/heapro/daz053

Cummins, S. (2007). Commentary: Investigating neighbourhood effects on health – Avoiding the "Local Trap". *International Journal of Epidemiology, 36*(2), 355–357. https://doi.org/10.1093/ije/dym033

Cummins, S., Curtis, S., Diez-Roux, A. V., & Macintyre, S. (2007). Understanding and representing 'place' in health research: A relational approach. *Social Science & Medicine, 65*(9), 1825–1838. https://doi.org/10.1016/j.socscimed.2007.05.036

Curtis, S., Southall, H., Congdon, P., & Dodgeon, B. (2004). Area effects on health variation over the life-course: Analysis of the longitudinal study sample in England using new data on area of residence in childhood. *Social Science & Medicine, 58*(1), 57–74. https://doi.org/10.1016/S0277-9536(03)00149-7

Cutchin, M. P. (2007). The need for the "new health geography" in epidemiologic studies of environment and health. *Health & Place, 13*(3), 725–742.

Cutchin, M. P., Eschbach, K., Mair, C. A., Ju, H., & Goodwin, J. S. (2011). The socio-spatial neighborhood estimation method: An approach to operationalizing the neighborhood concept. *Health & Place, 17*(5), 1113–1121. https://doi.org/10.1016/j.healthplace.2011.05.011

Darroch, F., & Giles, A. (2014). Decolonizing health research: Community-based participatory research and postcolonial feminist theory. *Canadian Journal of Action Research, 15*(3), 22–36.

Dennis, S. F., Gaulocher, S., Carpiano, R. M., & Brown, D. (2009). Participatory photo mapping (PPM): Exploring an integrated method for health and place research with young people. *Health & Place, 15*(2), 466–473. https://doi.org/10.1016/j.healthplace.2008.08.004

Diez Roux, A. V. (2004). The study of group-level factors in epidemiology: Rethinking variables, study designs, and analytical approaches. *Epidemiologic Reviews, 26*(1), 104–111. https://doi.org/10.1093/epirev/mxh006

Diez Roux, A. V., & Mair, C. (2010). Neighborhoods and health: Neighborhoods and health. *Annals of the New York Academy of Sciences, 1186*(1), 125–145. https://doi.org/10.1111/j.1749-6632.2009.05333.x

Fang, M. L., Woolrych, R., Sixsmith, J., Canham, S., Battersby, L., & Sixsmith, A. (2016). Place-making with older persons: Establishing sense-of-place through participatory community mapping workshops. *Social Science & Medicine, 168*, 223–229. https://doi.org/10.1016/j.socscimed.2016.07.007

Frumkin, H. (2006). The measure of place. *American Journal of Preventive Medicine, 31*(6), 530–532.

Galster, G., & Sharkey, P. (2017). Spatial foundations of inequality: A conceptual model and empirical overview. *RSF: The Russell Sage Foundation Journal of the Social Sciences, 3*(2), 1. https://doi.org/10.7758/rsf.2017.3.2.01

Gustafsson, P. E., San Sebastian, M., Janlert, U., Theorell, T., Westerlund, H., & Hammarström, A. (2014). Life-course accumulation of neighborhood disadvantage and allostatic load: Empirical integration of three social determinants of health frameworks. *American Journal of Public Health, 104*(5), 904–910.

Hägerstrand, T. (1970). What about people in regional science? *Papers of the Regional Science Association, 24*(1), 6–21. https://doi.org/10.1007/BF01936872

Hand, C. L., Rudman, D. L., Huot, S., Gilliland, J. A., & Pack, R. L. (2018). Toward understanding person–place transactions in neighborhoods: A qualitative-participatory geospatial approach. *The Gerontologist, 58*(1), 89–100. https://doi.org/10.1093/geront/gnx064

Hertzman, C., & Power, C. (2003). Health and human development: Understandings from life-course research. *Developmental Neuropsychology, 24*(2–3), 719–744. https://doi.org/10.1080/87565641.2003.9651917

Hertzman, C., Power, C., Matthews, S., & Manor, O. (2001). Using an interactive framework of society and lifecourse to explain self-rated health in early adulthood. *Social Science & Medicine, 53*(12), 1575–1585.

Israel, B. A., Schulz, A. J., Parker, E. A., & Becker, A. B. (1998). Review of community-based research: Assessing partnership approaches to improve public health. *Annual Review of Public Health, 19*(1), 173–202.

Israel, B. A., Coombe, C. M., Cheezum, R. R., Schulz, A. J., McGranaghan, R. J., Lichtenstein, R., Reyes, A. G., Clement, J., & Burris, A. (2010). Community-based participatory research: A capacity-building approach for policy advocacy aimed at eliminating health disparities. *American Journal of Public Health, 100*(11), 2094–2102. https://doi.org/10.2105/AJPH.2009.170506

John, D. H., & Gunter, K. (2016). engAGE in community: Using mixed methods to mobilize older people to elucidate the age-friendly attributes of urban and rural places. *Journal of Applied Gerontology, 35*(10), 1095–1120. https://doi.org/10.1177/0733464814566679

Jones, M., & Pebley, A. R. (2014). Redefining neighborhoods using common destinations: Social characteristics of activity spaces and home census tracts compared. *Demography, 51*(3), 727–752. https://doi.org/10.1007/s13524-014-0283-z

Kemp, S. P. (2011). Place, history, memory: Thinking time within place. In L. M. Burton, S. A. Matthews, M. Leung, S. P. Kemp, & D. T. Takeuchi (Eds.), *Communities, neighborhoods, and health* (pp. 3–19). Springer. https://doi.org/10.1007/978-1-4419-7482-2_1

Kwan, M.-P. (2009). From place-based to people-based exposure measures. *Social Science & Medicine, 69*(9), 1311–1313. https://doi.org/10.1016/j.socscimed.2009.07.013

Lantz, P., Israel, B. A., Schulz, A. J., & Reyes, A. G. (2006). Community-based participatory research: Rationale and relevance for social epidemiology. In J. M. Oakes & J. Kaufman (Eds.), *Methods in social epidemiology* (pp. 229–266). Jossey-Bass.

Leung, M. W. (2004). Community based participatory research: A promising approach for increasing epidemiology's relevance in the 21st century. *International Journal of Epidemiology, 33*(3), 499–506. https://doi.org/10.1093/ije/dyh010

Lin, L., & Moudon, A. V. (2010). Objective versus subjective measures of the built environment, which are most effective in capturing associations with walking? *Health & Place, 16*(2), 339–348. https://doi.org/10.1016/j.healthplace.2009.11.002

Lynch, J., & Smith, G. D. (2005). A life course approach to chronic disease epidemiology. *Annual Review of Public Health, 26*(1), 1–35. https://doi.org/10.1146/annurev.publhealth.26.021304.144505

Macintyre, S., Ellaway, A., & Cummins, S. (2002). Place effects on health: How can we conceptualise, operationalise and measure them? *Social Science & Medicine, 55*(1), 125–139.

Matthews, S. A. (2008). The salience of neighborhood. *American Journal of Preventive Medicine, 34*(3), 257–259. https://doi.org/10.1016/j.amepre.2007.12.001

Matthews, S. A. (2011). Spatial polygamy and the heterogeneity of place: Studying people and place via egocentric methods. In L. M. Burton, S. A. Matthews, M. Leung, S. P. Kemp, & D. T. Takeuchi (Eds.), *Communities, neighborhoods, and health* (pp. 35–55). Springer. https://doi.org/10.1007/978-1-4419-7482-2_3

Matthews, S. A., & Yang, T.-C. (2013). Spatial polygamy and contextual exposures (SPACEs): Promoting activity space approaches in research on place and health. *American Behavioral Scientist, 57*(8), 1057–1081. https://doi.org/10.1177/0002764213487345

Matthews, S. A., Detwiler, J. E., & Burton, L. M. (2005). Geo-ethnography: Coupling geographic information analysis techniques with ethnographic methods in urban research. *Cartographica: The International Journal for Geographic Information and Geovisualization, 40*(4), 75–90. https://doi.org/10.3138/2288-1450-W061-R664

McGibbon, E., Mulaudzi, F. M., Didham, P., Barton, S., & Sochan, A. (2014). Toward decolonizing nursing: The colonization of nursing and strategies for increasing the counter-narrative. *Nursing Inquiry, 21*(3), 179–191. https://doi.org/10.1111/nin.12042

Merlo, J. (2011). Contextual influences on the individual life course: Building a research framework for social epidemiology. *Psychosocial Intervention, 20*(1), 109–118.

Milton, S., Pliakas, T., Hawkesworth, S., Nanchahal, K., Grundy, C., Amuzu, A., Casas, J.-P., & Lock, K. (2015). A qualitative geographical information systems approach to explore how older people over 70 years interact with and define their neighbourhood environment. *Health & Place, 36*, 127–133. https://doi.org/10.1016/j.healthplace.2015.10.002

Minkler, M. (2000). Using participatory action research to build healthy communities. *Public Health Reports (Washington, D.C.: 1974), 115*(2–3), 191–197.

Minkler, M. (2010). Linking science and policy through community-based participatory research to study and address health disparities. *American Journal of Public Health, 100*(S1), S81–S87.

Morello-Frosch, R., Jr., Pastor, M., Sadd, J., Porras, C., & Prichard, M. (2005). Citizens, science, and data judo: Leveraging secondary data analysis to build a community-academic collaborative for environmental justice in Southern California. In *Methods in community-based participatory research for health* (p. 22). Jossey-Bass.

Mujahid, M. S., Diez Roux, A. V., Morenoff, J. D., & Raghunathan, T. (2007). Assessing the measurement properties of neighborhood scales: From psychometrics to ecometrics. *American Journal of Epidemiology, 165*(8), 858–867. https://doi.org/10.1093/aje/kwm040

Mundel, E., & Chapman, G. E. (2010). A decolonizing approach to health promotion in Canada: The case of the Urban Aboriginal Community Kitchen Garden Project. *Health Promotion International, 25*(2), 166–173. https://doi.org/10.1093/heapro/daq016

Nazmi, A., Diez Roux, A., Ranjit, N., Seeman, T. E., & Jenny, N. S. (2010). Cross-sectional and longitudinal associations of neighborhood characteristics with inflammatory markers: Findings from the multi-ethnic study of atherosclerosis. *Health & Place, 16*(6), 1104–1112. https://doi.org/10.1016/j.healthplace.2010.07.001

Perchoux, C., Chaix, B., Cummins, S., & Kestens, Y. (2013). Conceptualization and measurement of environmental exposure in epidemiology: Accounting for activity space related to daily mobility. *Health & Place, 21*, 86–93. https://doi.org/10.1016/j.healthplace.2013.01.005

Petteway, R., Mujahid, M., & Allen, A. (2019a). Understanding embodiment in place-health research: Approaches, limitations, and opportunities. *Journal of Urban Health*. https://doi.org/10.1007/s11524-018-00336-y

Petteway, R., Mujahid, M., Allen, A., & Morello-Frosch, R. (2019b). Towards a people's social epidemiology: Envisioning a more inclusive and equitable future for social epi research and practice in the 21st century. *International Journal of Environmental Research and Public Health, 16*(20), 3983. https://doi.org/10.3390/ijerph16203983

Pruitt, S. L., Jeffe, D. B., Yan, Y., & Schootman, M. (2012). Reliability of perceived neighbourhood conditions and the effects of measurement error on self-rated health across urban and rural neighbourhoods. *Journal of Epidemiology and Community Health, 66*(4), 342–351. https://doi.org/10.1136/jech.2009.103325

Rainham, D., McDowell, I., Krewski, D., & Sawada, M. (2010). Conceptualizing the healthscape: Contributions of time geography, location technologies and spatial ecology to place and health research. *Social Science & Medicine, 70*(5), 668–676. https://doi.org/10.1016/j.socscimed.2009.10.035

Richardson, D. M., & Nuru-Jeter, A. M. (2012). Neighborhood contexts experienced by young Mexican-American women: Enhancing our understanding of risk for early childbearing. *Journal of Urban Health, 89*(1), 59–73. https://doi.org/10.1007/s11524-011-9627-9

Riva, M., Apparicio, P., Gauvin, L., & Brodeur, J.-M. (2008). Establishing the soundness of administrative spatial units for operationalising the active living potential of residential environments: An exemplar for designing optimal zones. *International Journal of Health Geographics, 7*(1), 43. https://doi.org/10.1186/1476-072X-7-43

Santos, S. M., Chor, D., Werneck, G. L., & Coutinho, E. S. F. (2007). Associação entre fatores contextuais e auto-avaliação de saúde: Uma revisão sistemática de estudos multinível. *Cadernos de Saúde Pública, 23*, 2533–2554.

Schultz, A., Zenk, S., Kannan, S., Koch, M., Israel, B., & Stokes, C. (2005). Community-based participatory approach to survey design and implementation: The Healthy Environments Community Survey. In B. Israel, E. A. Parker, E. Eng, & A. Schulz (Eds.), *Methods for conducting community-based participatory research for health* (pp. 107–127). Jossey-Bass.

Schulz, A. J., Mentz, G., Lachance, L., Zenk, S. N., Johnson, J., Stokes, C., & Mandell, R. (2013). Do observed or perceived characteristics of the neighborhood environment mediate associations between neighborhood poverty and cumulative biological risk? *Health & Place, 24*, 147–156. https://doi.org/10.1016/j.healthplace.2013.09.005

Sharkey, P., & Elwert, F. (2011). The legacy of disadvantage: Multigenerational neighborhood effects on cognitive ability. *American Journal of Sociology, 116*(6), 1934–1981.

Spielman, S. E., & Yoo, E. (2009). The spatial dimensions of neighborhood effects. *Social Science & Medicine, 68*(6), 1098–1105. https://doi.org/10.1016/j.socscimed.2008.12.048

Tong, C., Sims-Gould, J., & McKay, H. (2016). InterACTIVE Interpreted Interviews (I3): A multilingual, mobile method to examine the neighbourhood environment with older adults. *Social Science & Medicine, 168*, 207–213. https://doi.org/10.1016/j.socscimed.2016.08.010

van Deurzen, I., Rod, N. H., Christensen, U., Hansen, Å. M., Lund, R., & Dich, N. (2016). Neighborhood perceptions and allostatic load: Evidence from Denmark. *Health & Place, 40*, 1–8. https://doi.org/10.1016/j.healthplace.2016.04.010

Villanueva, K., Giles-Corti, B., Bulsara, M., McCormack, G. R., Timperio, A., Middleton, N., Beesley, B., & Trapp, G. (2012). How far do children travel from their homes? Exploring children's activity spaces in their neighborhood. *Health & Place, 18*(2), 263–273. https://doi.org/10.1016/j.healthplace.2011.09.019

Wallerstein, N., & Duran, B. (2010). Community-based participatory research contributions to intervention research: The intersection of science and practice to improve health equity. *American Journal of Public Health, 100*(S1), S40–S46. https://doi.org/10.2105/AJPH.2009.184036

Wallerstein, N. B., Yen, I. H., & Syme, S. L. (2011). Integration of social epidemiology and community-engaged interventions to improve health equity. *American Journal of Public Health, 101*(5), 822–830. https://doi.org/10.2105/AJPH.2008.140988

Weden, M. M., Carpiano, R. M., & Robert, S. A. (2008). Subjective and objective neighborhood characteristics and adult health. *Social Science & Medicine, 66*(6), 1256–1270. https://doi.org/10.1016/j.socscimed.2007.11.041

Wen, M., Hawkley, L. C., & Cacioppo, J. T. (2006). Objective and perceived neighborhood environment, individual SES and psychosocial factors, and self-rated health: An analysis of older adults in Cook County, Illinois. *Social Science & Medicine, 63*(10), 2575–2590. https://doi.org/10.1016/j.socscimed.2006.06.025

York Cornwell, E., & Cagney, K. A. (2017). Aging in activity space: Results from smartphone-based GPS-tracking of urban seniors. *The Journals of Gerontology: Series B, 72*(5), 864–875. https://doi.org/10.1093/geronb/gbx063

Chapter 2
Spatial Knowledge, Representation, + Place-Health Narratives: Youth Photovoice Perspectives on a "Food Desert"

(Mis)Representation

A Baltimore "Food Desert" Story

You live in a food desert.

Those words are straight from a flyer (Fig. 2.1) my colleagues and I passed out to residents as part of our efforts to develop a "virtual supermarket" program within various Baltimore City neighborhoods (Aggarwal & Petteway, 2010; Lagisetty et al., 2017). Of course, critiques of this phrasing abound now (Bedore, 2013; Hill, 2017; Howerton & Trauger, 2017; Sadler et al., 2016a; Taylor & Ard, 2015), and there is a growing body of work, e.g., radical and Black food geographies, more appropriately focused on historic and present structural inequities of power related to food justice and sovereignty (Bradley & Herrera, 2016; Brown et al., 2020; Glennie & Alkon, 2018; Hammelman et al., 2020; Horst et al., 2017; LaDuke, 2019; Levkoe et al., 2020; Ramírez, 2015; Reese, 2019; Trauger, 2017). I found the phrasing disturbingly problematic myself at the time. Yet the decision to use those specific words was less a decision than a compulsion—local health jurisdictions and the public health field at-large were increasingly obsessed with this framing—it was trending, it had followers, it was magical. Moreover, in Baltimore, it was a leading frame for the city's overall food policy strategy (BCFPTF, 2009)—one which, in practice, seemed to suggest that people actually lived in food deserts, as opposed to contested, unevenly developed geographic spaces structured by racial and economic segregation that (re)produce inequitable environments over multiple generations (Bower et al., 2014; Brown, 2015, 2021; Deener, 2017; Glotzer, 2020; Gomez, 2015; Pietila, 2010). The food desert frame, in effect, was that little memory-erasing thing from *Men in Black*: look here…

Oh, look, a food desert! I hadn't noticed it before—wonder how long it's been here. Does the temperature drop at night like in the ones out West? Does it rain enough to grow kale?

© Springer Nature Switzerland AG 2022

R. J. Petteway, *Representation, Re-Presentation, and Resistance*, Global Perspectives on Health Geography, https://doi.org/10.1007/978-3-031-06141-7_2

Fig. 2.1 Virtual supermarket community outreach flyer, 2010. Example outreach flyer we developed for drawing potential customers to the virtual supermarket program. Notice the "food desert" framing, as well as the emphasis on geographic distance

Redlining and racist zoning (Brown, 2021; Joint Center, 2012; Power, 1983)? Hot sand. Urban Renewal (Blessett, 2020; Heyda, 2020)? A rattlesnake. HOPE VI (Roche Jr., 2002)? A scorpion. A highway to nowhere (Brown, 2015; Gioielli, 2011)? Tumbleweed. The largest property owner in the city not paying fair property taxes and actively displacing residents (Gomez, 2015; National Nurses United, 2019; Richman, 2019)? A Gila monster.

Alas, we passed out these flyers, complete with a map of their "neighborhood" as defined by, well, us, of course. As part of my role as (social) epidemiologist for the local health department, I analyzed, geocoded, and mapped dozens of community health indicators in developing Neighborhood Health Profiles (Petteway & Ames, 2011). These included data on health outcomes (e.g., mortality rates for various causes, life expectancy), as well as data on social determinants of health (e.g., food environment, housing conditions, built environment), along with population demos (e.g., race, household income, educational attainment, sex). Among the many maps I produced were maps of the food environment for all of Baltimore's 55 community statistical areas (see Fig. 2.2, for example), or CSAs, based on residential census tracts.

The virtual supermarket work, as well as the mapping and community assessment work, coalesced with and in many ways advanced the work of the Baltimore City Food Policy Taskforce, which put forth a set of recommendations in December 2009 to guide the City's food policy efforts (BCFPTF, 2009). At the time, efforts to address "childhood obesity" and diet-related causes of death and morbidity within the City had gradually expanded to encompass underlying social determinants of health, particularly inequities in access to healthy food options, with critical work led by researchers at Johns Hopkins Center for a Livable Future (Franco et al., 2008, 2009; Santo et al., 2014). Accordingly, community food environment was increasingly being recognized as a fundamental determinant of dietary/purchasing behaviors and diet-related health inequities.

Unfortunately, efforts were led primarily by "outsiders," e.g., academics, administrators, and epidemiologists such as myself, and not the actual community residents. For example, the Taskforce itself included 18 members, many of whom were not from Baltimore City, did not live in Baltimore City, and/or were not from the communities identified as food deserts. Moreover, there was no youth presence on the Taskforce—despite the significance of lifecourse considerations for diet-related health chronic disease outcomes (Darnton-Hill et al., 2004; Devine, 2005; Lynch & Smith, 2005; Wethington & Johnson-Askew, 2009), and despite the fact that 2 of the 10 priority areas directly affected/implicated youth. Moreover, local efforts and discussions had proceeded with a robustly ahistorical and apolitical lens. For example, "racism" and "segregation" were not mentioned a single time in the Taskforce report, nor were pertinent matters related to spatial dispossession and displacement from urban renewal and community development processes, many of which were propagated by the institutions that employed at least four members on the Taskforce.

In this context, in more ways than one, the maps I and others produced to represent residents' food geographies supported the creation/maintenance of spatial narratives of absence and pathology—maps produced largely in service of/deference to dominant positivist and colonial epistemologies within public health knowledge production. While these maps were not deliberate acts of knowledge and political erasure, they had the effect of masking resident voice and agency, and severing present conditions from historic contexts—that is, they were not only "colorblind" but power-blind. As such, the narrative surrounding Baltimore's food environments, and specifically the "food desert" frame, of course failed to capture the nuanced and lived reality of many—if not most—residents in those so-labeled neighborhoods. And it rendered invisible the myriad historic and contemporary processes and

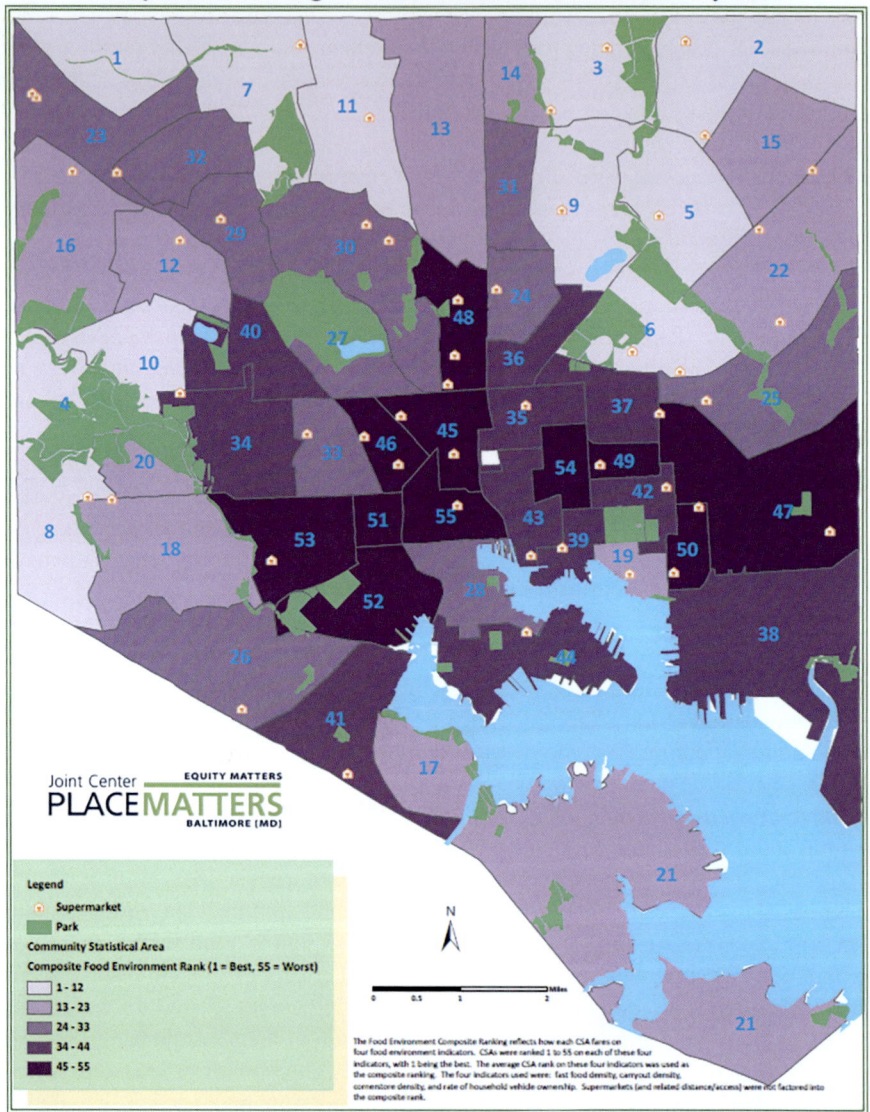

Fig. 2.2 Baltimore city food environment ranking by Community Statistical Area (CSA), 2010. Map I produced showing composite ranking of food environments based on community statistical areas, or CSAs. CSAs consist of groupings of census tracts. The measures factored into the ranking include: fast food density, cornerstore density, carry-out density, household vehicle ownership rates, and average travel time to nearest supermarket from the "centroid" of the most populous census track of each CSA. The orange symbols represent supermarkets. The darker the purple, the lower the quality of the food environment

mechanisms that created and function to maintain inequities in neighborhood food environments that disproportionately—by design—landed upon Black and Brown bodies.

The following sections present and discuss one effort to nuance this narrative of "food deserts" from the perspective of youth residents of one so-defined neighborhood, using photovoice as a process to re-present their geographies of food and health opportunity/inequity.

Re-Presentation

The Project Food for Thought Photovoice Project

Background

I developed the "Project Food for Thought" youth photovoice project as a collaboration with a local art college and MFA student, Lauren Adams. We partnered with the Baltimore Collaborative for Health Equity (formerly Place Matters) team, Equity Matters, on various aspects of the project. The project had a particular focus on food environment inequities as a barrier to community health in two East Baltimore neighborhoods that had been identified as part of a food desert. The spatial location of the health department was of particular importance for this project—itself being located in the same East Baltimore food desert. Moreover, it was essentially surrounded by public housing communities, with which I had worked with/within intermittently during my time as a social epidemiologist, establishing some baseline level of rapport and presence. As a former public housing resident myself, and given the evidence regarding health inequities among public housing communities (Digenis-Bury et al., 2008; Harris & Kaye, 2004; Howell et al., 2005; Keene & Geronimus, 2011; Manjarrez et al., 2007), I was especially interested in engaging with residents in a more formal and sustained capacity. But despite the close proximity of these communities, there were no broader, long-term community engagement efforts within the department to assess and respond to their health concerns. This project was one way to open up this sort of collaborative relationship.

Generally, we developed Project Food for Thought for two core reasons: (1) create an opportunity to include youth perspectives within the food environment narrative, and (2) deepen understanding of food environment concerns, priorities, and opportunities by including peoples' lived experience as part of the data story. Basically, too much of the conversation and narrative was created and driven by folks who had never even stepped foot in one of these food deserts, let alone had any significant connection to/history with the people residing in them. Thus, this particular project was in many ways a deliberate effort to counter and add context to the pervading narratives at the time—ones which tended to focus on disembodied health behaviors, mask resident agency, and obscure mechanisms/pathways of social and political accountability.

In this context, photovoice presented as an ideal community assessment method. Generally, this method entails participants taking photos in a social-documentary style on a posed topic, generating narratives (written) for each photo, and discussing photos in groups to identify salient themes. With conceptual roots in popular education and education for critical consciousness (Freire, 2000), feminist notions of situated and partial knowledge (Haraway, 1988; McDowell, 1992), and other participatory frameworks and photo-elicitation traditions (Catalani & Minkler, 2010; Wang & Burris, 1997), photovoice has been used for years for community assessment projects with youth (Morales-Campos et al., 2015; Petteway et al., 2019; Richardson & Nuru-Jeter, 2012; Strack et al., 2004; Wang, 2006). The method is premised on three core goals of the research process (Wang & Burris, 1997, p. 369):

> (1) to enable people to record and reflect their community's strengths and concerns, (2) to promote critical dialogue and knowledge about important issues through large and small group discussion of photographs, and (3) to reach policymakers/stakeholders.

Photovoice, accordingly, was well-suited for this project's intention, drawing strength from its ability to elicit contextualized meanings narrated in the voice and words of participants. For this project, youth used photography to visually document their daily/weekly experiences with/perceptions of their community food environment. We viewed this as an opportunity for youth to visually examine and (re)present their lived experiences within a "food desert," and thereby nuance existing narratives and articulate counterstories that could inform City efforts.

Photovoice Recruitment and Training

A total of eight middle-school-aged youth joined the project. We recruited participants from three public housing communities in East Baltimore: Douglass Homes, Pleasant View Gardens, and Perkins Homes. These communities were located in a predominantly Black (88%), high-poverty (44%) neighborhood in East Baltimore that was home to 40 carryouts, 6 fast food restaurants, 15 cornerstores, and 0 supermarkets at the time (Petteway & Adams, 2011; Petteway & Ames, 2011). The nearest supermarket was over 1 mile away and 68% of households did not own a vehicle. All recruitment activities were co-led with staff at the Carmelo Anthony Youth Development Center (CAYDC). The CAYDC was right across the street from the health department and the site of all photovoice project meetings, with the exception of a final review/discussion meeting at the health department and the final project photo exhibit held at City Hall.

The youth attended three Orientation and Training Sessions where we discussed participation questions (e.g., parent permission, assent), community health and food environment basics, as well as an overview of participatory research principles and ethics. We also trained participants in basic camera use and general ethics/guiding principles of photography (Photo 2.1), as is standard for photovoice projects. We provided the youth with digital cameras, which they were able to keep upon project

Photo 2.1 Project Food for Thought participants during photography training session. Photo of youth researchers learning to use their digital cameras as part of the photovoice training

Table 2.1 The PHOTO photovoice narrative guide

P	describe your **P**icture
H	what is **H**appening in your picture?
O	why did you take a picture **O**f this?
T	what does this picture **T**ell us about life in your community?
O	how does this picture provide **O**pportunities for us to improve life in your community?

Summary of photovoice guide. Used for this project as presented in Pies and Parathasarthy (2008), developed based on their community-engaged work within Contra Costa Health Services' Family, Maternal, and Child Health Program

completion. They took photos on their own time in between project meetings as they went about their daily activities.

Photo Review and Discussion

We met with youth at regular intervals (weekly, generally) to facilitate discussion of their photos and guide them in creating photo narratives, using the *PHOTO* photovoice narrative guide containing five critical inductive questions (Table 2.1). We used this guide because it was shorter and used simpler language/syntax than the more commonly used *SHOWeD* guide (Wang, 1999). Youth attended eight photo review sessions to discuss their photos and complete worksheets with the PHOTO guide. We did not audio record these sessions, but took handwritten notes during photo discussions. Youth selected their top five favorite photos for printing and/or for turning into postcards. At the final review session, we met with youth at the health department to frame their printed photos, review and discuss their framed photos, and organize them into broad thematic groups. Youth also completed a brief project "evaluation" worksheet to reflect on their experience (see below).

Photovoice Findings: Photo Themes

Through review, discussion, and the participatory sorting of their framed photos, the youth identified four broad themes: (1) food environment, (2) cornerstore snacks, (3) healthy options, and (4) frozen and processed foods. The illustrative photos below are included because they were the most salient within each respective thematic area, based on the discussion each elicited during review sessions, i.e., these photos generated the most conversation and reflection, and/or because the youth chose to have them framed for the photovoice exhibit or printed as postcards (see below). Photo discussion excerpts are shown where available.

Theme #1: Food Environment (Photo 2.2)

Photographer narrative:

> "I took a picture of this because it's food-related. I don't like to go there. It's right around the corner from my house."

Photo 2.2 "Sun Grocery"

Discussion excerpts:

> Youth 1: "it says WIC and Food Stamps… looks old and rundown… I stop there to get my snacks…"
> Youth 2: "my cousin has to shop there"
> Youth 3: "the meat is nasty"
> [when asked about if they have fruits and veggies]
> Youth 4: "they have fruit cups"
> [when prompted to think about how to change the store]
> Youth: 2: "would get fresh everything if I could change it"
> Youth 5: "only one aisle of junk food… and put the food pyramid at each aisle, with fruits and veggies"
> Photographer: "would want the glass so we wouldn't get robbed"

Theme #2: Cornerstore Snacks (Photo 2.3)

Photographer narrative:

> "Hot Cheetos are my absolute favorite snack. I can get this product not too far from my house."

Discussion excerpt:

> Youth 1: "is hot stuff good for you?"
> Photographer: "I eat them every day"
> Youth 2: "do chips expire?"
> Youth 3: "yeah, they go stale after a while"
> [when prompted to compare expiration of chips vs. a banana]
> Youth 3: "chips last 2, 3 months… bananas only last a week"
> Youth 1: "things with longer shelf life has a lot of chemicals"

Theme #3: Healthy Options (Photo 2.4)

Photographer narrative:

> "I took a picture because I think bananas are very delicious and good for you. You can get them from cornerstores."

Theme #4: Frozen and Processed Foods (Photo 2.5)

Photographer narrative:

> "The day my mother went shopping and we had a lot of food. So much food that I can't explain.
> But I'm only thinking about the pickles because it's close to cucumbers, and they are healthy."

Discussion excerpt:

> Youth 1: "my mom shops and it lasts for a month"
> Youth 2: "in my photos, all the food is from a factory"

Photo 2.3 "Hot Cheetos"

Photo 2.4 "Bananas"

Photovoice Findings: Evaluation Worksheets

As part of the photovoice process, youth completed brief "evaluation" worksheets that asked them to reflect on their experience. Figure 2.3 shows illustrative examples of youth responses to the following three prompts/questions:

1. Name one thing you learned about photography.
2. Is there anything you wish you could buy at cornerstores? What is it?
3. Name three things that, if changed, could help your community have easier access to better food.

Photo 2.5 "Shopping"

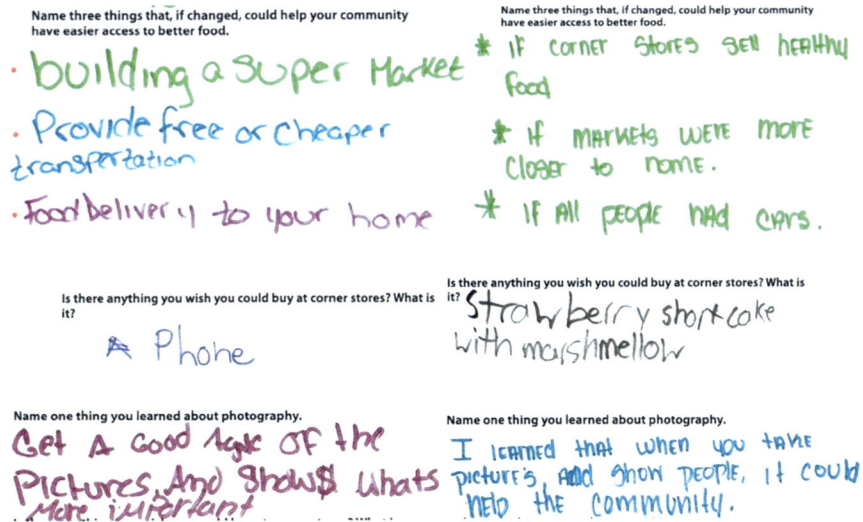

Fig. 2.3 Example youth photovoice evaluation worksheets. Example quotes from youth evaluations of the photovoice project

City Hall Photovoice Exhibit and Postcards

Youth's framed photos and accompanying narratives were used to host a project exhibit at City Hall. Lauren Adams worked with the youth to design and plan the exhibit, incorporating their photos and narratives on display boards within the central rotunda (Photo 2.6). We hosted a formal exhibit event for youth to guide attendees through their work and discuss their ideas for how to improve their community food environment. The event was attended by City Council members, health department leadership, and various other community members and organization leaders.

In a discussion to generate ideas for other potential activities/actions to continue their engagement, the youth elected to create postcards out of a selection of their photos (Fig. 2.4). The youth worked with Lauren to finalize their photo choices and provide feedback as she developed them. Each youth participant was provided a stack of postcards to disseminate within their community. Additional postcards were distributed by the Baltimore Place Matters team and at Lauren's MFA art

Photo 2.6 Project Food for Thought photo exhibit at City Hall. Photo from project exhibit in City Hall. Youth photos were displayed in the main rotunda, where they had a chance to meet and discuss their work with various City officials

Food for Thought 2011: Youth Perspectives on a Baltimore Food Desert
Empowering Youth through Photovoice

Students documented their daily experiences living in a food desert—a neighborhood that lacks sufficient access to healthy options. For more information or to get involved, visit *equity-matters.org* or email *info@equity-matters.org*.

COLLABORATORS | *Laura P. Adams* of the MICA Center for Design Practice, Ryan J. Petteway of the Baltimore City Health Department, Living Classrooms Foundation and Carmelo Anthony Youth Development Center, & Equity Matters

PHOTOGRAPHER | *Shatia, Grade 6*

Collaborative Social Design | A GRAPHIC DESIGN MFA THESIS BY LAUREN P. ADAMS | laurenpadams.com

Fig. 2.4 Example postcard created from youth photovoice photos. Example postcard created from youth photovoice photos. This one was entitled, "We Need Gardens,", and had the following photo narrative: "This place is by my school. It has so much grass that it could be a grassland. So I thought it would be a good place for [growing] fruits and vegetables"

thesis exhibit. The initial plan was to pre-address each postcard with contact info for the City Councilmember whose jurisdiction covered the youth's "food desert." However, as discussed below, this proved to be less than straightforward, and only a portion the postcards were pre-addressed in the intended manner.

Resistance

Discussion: Participatory Food Geographies as Resistance

The goal of presenting the work described in this chapter was to raise conceptual and analytical questions regarding the role of administrative agencies in mapping geographies of health—here, in relation to food geographies—and to highlight the value of community knowledge in (re)defining these geographies and narratives thereof. Moreover, I also wanted to touch on the potential consequences of narrative exclusion when visually and politically rendering these geographies, and potential conflicts between—and affordances of—various methods and forms of spatial representation within local health geographies. In this regard, the youth's work suggests perhaps three major takeaways and points for consideration.

First, the youth's photos and photo themes indicated that they were quite aware of the "food desert" status of their community. However, while theme #1 in many ways aligned with and reenforced the City's pre-existing narrative frame of "food desert," themes #2 and #4—while certainly not mutually exclusive of theme #1 (e.g., eating processed food because that's what is available in a low-quality food environment)—presented new perspectives for consideration. For example, for theme #2, it became apparent that frequent unhealthy snacks were a salient issue for the youth. This was clearly tied to the food desert frame captured via theme #1, but what was missing from this frame alone was this: the youth discussed frequently buying unhealthy snacks *on the way to and from school*. In other words, it wasn't just a matter of a generic census-tract-defined geographic area being declared a food desert, but was specifically tied to their daily mobility patterns between home and school. In this way, the youth's work reflects the importance of activity space approaches to understanding people's place-based health-related exposures (Kwan, 2009; Perchoux et al., 2013). Research published after the completion of this project has highlighted the conceptual and empirical value of such approaches (Browning & Soller, 2014; Cagney et al., 2020; Lipperman-Kreda et al., 2014; Villanueva et al., 2012), including as related to diet and food environments (Crawford et al., 2014; Horner & Wood, 2014; Kestens et al., 2010; Raskind et al., 2020; Sadler et al., 2016b; Widener et al., 2018; Zenk et al., 2011). Unfortunately, even while we (the health department, Center for a Livable Future, the Taskforce) had ample (even groundbreaking) geocoded food environment data available to us, ascertaining how these data aligned with youth's home-school mobility patterns was never considered. A 6th-grader's photo of *Flaming Hot Cheetos* is all it could have taken to nuance the discourse and epistemic orientation of the 18-member Taskforce. Of course, in retrospect, it's hard to determine how or if more deliberate, critical engagement of youth would have changed things; but participatory youth-centered work focused on food environments completed since then suggests a missed opportunity (Akom et al., 2016; Leung et al., 2017).

Second, and relatedly, youth traversed their routes to/from school in part because existing City policy made it such that they did not qualify for free bus fare. Baltimore City Public Schools did not have their own buses for middle and high school students, and students accordingly had to rely on the regional public transit service provided by the Maryland Transit Administration. School policy at the time stipulated that middle school children were not eligible for free bus fare unless their school was over 1.5 miles from their home, meaning that most students had to pay or walk. The generic food desert frame ignores how a core aspect of the youth's daily food-related exposures and behaviors were systematically created and maintained by ostensibly non-food-related factors. This perspective, introduced by the youth, suggested at least two core considerations by way of intervention: (1) systematically account for food environments along primary/major school walking routes and prioritize their mitigation (vs. a generic, geographically diffuse approach), and (2) revisit the policy regarding how bus pass availability is determined, such that youth have less exposure to the unhealthy food environments that make up their daily walk. Consideration of the first point aligned with—and could have refined— recommendation #7 of the Taskforce report, which focused on "improving the food environment around schools and recreation centers" (BCFPTF, 2009), as well as recommendation #9, which focused on using incentives and land-use zoning to create healthy food environments. Had youth been included in the development of the report, they could have offered guidance on (re)defining and more thoroughly specifying the spatial bounds of "the environment around schools recreation centers"— such that it was grounded in actual place-based exposures/experience of their daily food geographies. And perhaps, they could have collaboratively identified specific cornerstores to prioritize for efforts related to #9, like the healthy cornerstore initiative (Song et al., 2009; Wensel et al., 2019). Instead, youth knowledge and participation at the time was restricted to drawing pictures. Literally (BDP, 2016).

Third, as captured in theme #4 (Photo 2.5), it appeared that all of the youth were accustomed to one large-scale food purchasing trip to a grocery store each month. This is consistent with other research regarding food purchasing/security among low-income communities (Castner & Henke, 2011; Damon et al., 2013; Hamrick & Andrews, 2016). However, despite the fact that every single area of the city that was declared a "food desert," by definition, had high-poverty rates and low-vehicle ownership rates, matters related to food benefit distribution (i.e., supplemental nutrition assistance program, SNAP) and transportation equity did not crack the Taskforce's top ten list. As noted above, the City had a rather stingy approach to public bus fare for students. Meanwhile, the only free form of public transit at the time was the Charm City Circulator, which essentially catered to tourists, medical/university employees, and higher-income areas of the City—serving the so-called "White L" while averting the "Black Butterfly" (Brown, 2016; Brown, 2021). Had youth perspectives been centered—or at minimum, considered—their work suggests that critical discussions regarding how/when to distribute SNAP benefits and the necessity of prioritizing transportation would have been inevitable—even if this matter is beyond the scope of local policy. Instead, the youth and their families made the most of their inequitable food opportunities by purchasing foods that last (e.g., discussion excerpts for Photo 2.5 and Photo 2.2).

Overall, while youth identified some core concerns and introduced new perspectives to the food environment conversation, it remained unclear whom to bring their concerns to. As shown in Fig. 2.5, there were at least three City Council Members "responsible" for the conditions of their neighborhood food environments—throwing a bit of wrench into the postcard dissemination plan. And even though the youth had a chance to present and discuss their work with various councilmembers during the City Hall exhibit, they did not have the opportunity to more formally sit down with each of their specific three councilmembers. As a place-health scholar, this situation illuminates quite poignantly the limitations and fundamental flaws of standard local mapping/community assessment practices, in that they are based largely on imaginary geographies rooted in administrative bounds for which there is no direct link to political or jurisdictional accountability. Who is to be held accountable for a food desert based on census tract or ZIP code boundaries? Or when it spans three council districts?

At the time of completing this project, it was unclear whose desk to put this on, so to speak. And as a social epidemiologist and GIS-mapper at the time, I found it unsettling that much of the food geography work I had produced myself (and that produced by others, as well) was de facto apolitical and detached from core geographies of power and accountability. Thus, while the youth's work—as resistance and re-presentation—elucidated, at least partially, a few critical elements of their lived "food desert" geographies, it remained unclear how to center their voice in a meaningful way that could compel action in the political/policy sphere, i.e., legitimacy. As such, it was difficult to reconcile their work with and/or counter and destabilize already existing dominant spatial narratives of their community.

Conclusion

The work discussed here suggests a critical role for participatory and inclusive approaches in appropriately situating, appraising, and responding to spatial representations of urban food environment inequities. The youth's work revealed how participatory public health practice—here in relation to food environments—might inform policy and community change. Specifically, it suggests the potential value of photovoice as a standard component of public health/urban planning practice, as both process and method to re-present geographies of health and counter and/or enhance narratives thereof. Moreover, their work hints at what is often—through habit or design—left out the conversation. In Baltimore, these youth were excluded from contributing to discourse about their community food environment, effectively

Supermarkets and Corner/Grocery Stores by CSA

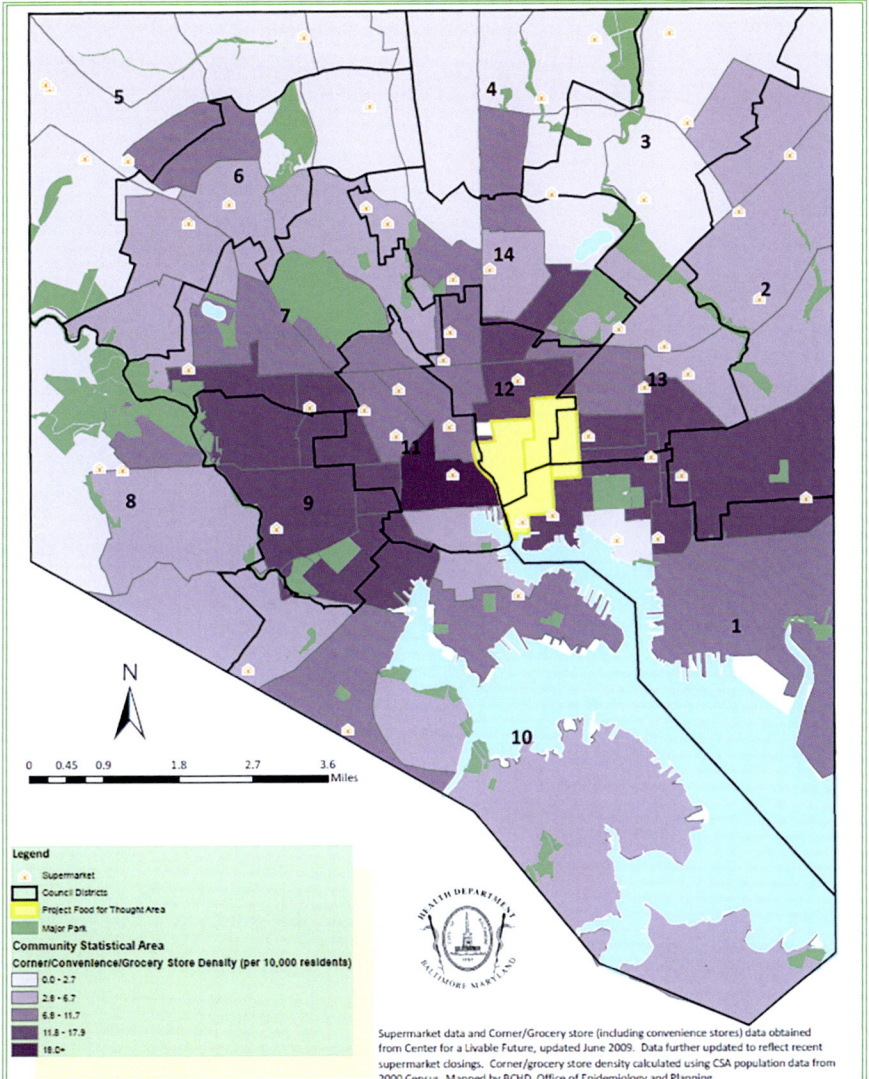

Fig. 2.5 Map of food for thought project area in relation to city council districts. Map showing composite ranking of food environments based on community statistical areas, or CSAs. CSAs consist of groupings of census tracts. The measures factored into the ranking include: fast food density, cornerstore density, carry-out density, household vehicle ownership rates, and average travel time to the nearest supermarket from the "centroid" of the most populous census track of each CSA. The orange symbols represent supermarkets. The darker the purple, the lower the quality of the food environment. The yellow shading represents the CSAs where the project took place. The black lines indicate City Councilmember jurisdictions

silenced and rendered invisible by dominant place knowledge production processes that afforded power of representation to credentialed "outsiders." This project, while by no means groundbreaking or powershifting in any material sense that I can speak to, served as an attempt to disrupt those processes, and in doing so, to resist efforts to erase youth's lived place knowledge as germane to food justice efforts.

References

Aggarwal, P., & Petteway, R. (2010). Virtual supermarket project: A fresh approach to addressing inequities in food access. *NACCHO annual 2010*.

Akom, A., Shah, A., Nakai, A., & Cruz, T. (2016). Youth participatory action research (YPAR) 2.0: How technological innovation and digital organizing sparked a food revolution in East Oakland. *International Journal of Qualitative Studies in Education, 29*(10), 1287–1307. https://doi.org/10.1080/09518398.2016.1201609

BCFPTF. (2009). *Baltimore City Food Policy Task Force: Final report and recommendations* (p. 39). Baltimore Department of Planning.

BDP. (2016, March 21). *Get fresh Baltimore*. Department of Planning. https://planning.baltimorecity.gov/baltimore-food-policy-initiative/get-fresh-baltimore

Bedore, M. (2013). Geographies of capital formation and rescaling: A historical-geographical approach to the food desert problem. *The Canadian Geographer/Le Géographe Canadien, 57*(2), 133–153. https://doi.org/10.1111/j.1541-0064.2012.00454.x

Blessett, B. (2020). Urban renewal and "Ghetto" development in Baltimore: Two sides of the same coin. *The American Review of Public Administration, 50*(8), 838–850. https://doi.org/10.1177/0275074020930358

Bower, K. M., Thorpe, R. J., Rohde, C., & Gaskin, D. J. (2014). The intersection of neighborhood racial segregation, poverty, and urbanicity and its impact on food store availability in the United States. *Preventive Medicine, 58*, 33–39. https://doi.org/10.1016/j.ypmed.2013.10.010

Bradley, K., & Herrera, H. (2016). Decolonizing food justice: Naming, resisting, and researching colonizing forces in the movement: Decolonizing food justice. *Antipode, 48*(1), 97–114. https://doi.org/10.1111/anti.12165

Brown, L. (2015). Down to the wire: Displacement and disinvestment in Baltimore city. In *State of Black Baltimore* (p. 18). Greater Baltimore Urban League. https://www.academia.edu/8619756/Down_to_the_Wire_Displacement_and_Disinvestment_in_Baltimore_City

Brown, L. (2016). Two Baltimores: The White L vs. the Black butterfly. *Baltimore Sun*. https://www.baltimoresun.com/citypaper/bcpnews-two-baltimores-the-white-l-vs-the-black-butterfly-20160628-htmlstory.html

Brown, L. T. (2021). *The Black butterfly: The harmful politics of race and space in America*. Johns Hopkins University Press. https://jhupbooks.press.jhu.edu/title/black-butterfly

Brown, J., Connell, K., Firth, J., & Hilton, T. (2020). The history of the land: A relational and place-based approach for teaching (more) radical food geographies. *Human Geography, 13*(3), 242–252. https://doi.org/10.1177/1942778620962024

Browning, C. R., & Soller, B. (2014). Moving beyond neighborhood: Activity spaces and ecological networks as contexts for youth development. *Cityscape (Washington, DC), 16*(1), 165.

Cagney, K. A., York Cornwell, E., Goldman, A. W., & Cai, L. (2020). Urban mobility and activity space. *Annual Review of Sociology, 46*(1), 623–648. https://doi.org/10.1146/annurev-soc-121919-054848

Castner, L., & Henke, J. (2011). *Benefit redemption patterns in the supplemental nutrition assistance program*. Food and Nutrition Service, Office of Research and Analysis. USDA. https://www.fns.usda.gov/snap/benefit-redemption-patterns-supplemental-nutrition-assistance-program

Catalani, C., & Minkler, M. (2010). Photovoice: A review of the literature in health and public health. *Health Education & Behavior, 37*(3), 424–451. https://doi.org/10.1177/1090198109342084

Crawford, T. W., Jilcott Pitts, S. B., McGuirt, J. T., Keyserling, T. C., & Ammerman, A. S. (2014). Conceptualizing and comparing neighborhood and activity space measures for food environment research. *Health & Place, 30*, 215–225. https://doi.org/10.1016/j.healthplace.2014.09.007

Damon, A. L., King, R. P., & Leibtag, E. (2013). First of the month effect: Does it apply across food retail channels? *Food Policy, 41*, 18–27. https://doi.org/10.1016/j.foodpol.2013.04.005

Darnton-Hill, I., Nishida, C., & James, W. P. T. (2004). A life course approach to diet, nutrition and the prevention of chronic diseases. *Public Health Nutrition, 7*(1a), 101–121. https://doi.org/10.1079/PHN2003584

Deener, A. (2017). The origins of the food desert: Urban inequality as infrastructural exclusion. *Social Forces, 95*(3), 1285–1309. https://doi.org/10.1093/sf/sox001

Devine, C. M. (2005). A life course perspective: Understanding food choices in time, social location, and history. *Journal of Nutrition Education and Behavior, 37*(3), 121–128. https://doi.org/10.1016/S1499-4046(06)60266-2

Digenis-Bury, E. C., Brooks, D. R., Chen, L., Ostrem, M., & Horsburgh, C. R. (2008). Use of a population-based survey to describe the health of boston public housing residents. *American Journal of Public Health, 98*(1), 85–91. https://doi.org/10.2105/AJPH.2006.094912

Franco, M., Diez Roux, A. V., Glass, T. A., Caballero, B., & Brancati, F. L. (2008). Neighborhood characteristics and availability of healthy foods in Baltimore. *American Journal of Preventive Medicine, 35*(6), 561–567. https://doi.org/10.1016/j.amepre.2008.07.003

Franco, M., Diez-Roux, A. V., Nettleton, J. A., Lazo, M., Brancati, F., Caballero, B., Glass, T., & Moore, L. V. (2009). Availability of healthy foods and dietary patterns: The multi-ethnic study of atherosclerosis. *The American Journal of Clinical Nutrition, 89*(3), 897–904. https://doi.org/10.3945/ajcn.2008.26434

Freire, P. (2000). *Pedagogy of the oppressed* (30th Anniversary ed.). Continuum.

Gioielli, R. (2011). "We must destroy you to save you": Highway construction and the city as a modern commons. *Radical History Review, 2011*(109), 62–82. https://doi.org/10.1215/01636545-2010-015

Glennie, C., & Alkon, A. H. (2018). Food justice: Cultivating the field. *Environmental Research Letters, 13*(7), 073003. https://doi.org/10.1088/1748-9326/aac4b2

Glotzer, P. (2020). *How the suburbs were segregated: Developers and the business of exclusionary housing, 1890–1960*. Columbia University Press.

Gomez, M. (2015). *Race, class, power, and organizing in East Baltimore: Rebuilding abandoned communities in America*. Rowman & Littlefield. https://rowman.com/ISBN/978-0-7391-7500-2

Hammelman, C., Reynolds, K., & Levkoe, C. Z. (2020). Toward a radical food geography praxis: Integrating theory, action, and geographic analysis in pursuit of more equitable and sustainable food systems. *Human Geography, 13*(3), 211–227. https://doi.org/10.1177/1942778620962034

Hamrick, K. S., & Andrews, M. (2016). SNAP participants' eating patterns over the benefit month: A time use perspective. *PLoS One, 11*(7), e0158422. https://doi.org/10.1371/journal.pone.0158422

Haraway, D. (1988). Situated knowledges: The science question in feminism and the privilege of partial perspective. *Feminist Studies, 14*(3), 575–599. https://doi.org/10.2307/3178066

Harris, L. E., & Kaye, D. R. (2004). How are HOPE VI families faring? *Health, 8.*

Heyda, P. (2020). Urban renewal and school reform in Baltimore: Rethinking the 21st century public school. *Journal of Urban Design, 25*(6), 812–816. https://doi.org/10.1080/13574809.2020.1814136

Hill, A. B. (2017). Critical inquiry into Detroit's "food desert" metaphor. *Food and Foodways, 25*(3), 228–246. https://doi.org/10.1080/07409710.2017.1348112

Horner, M. W., & Wood, B. S. (2014). Capturing individuals' food environments using flexible space-time accessibility measures. *Applied Geography, 51*, 99–107. https://doi.org/10.1016/j.apgeog.2014.03.007

Horst, M., McClintock, N., & Hoey, L. (2017). The intersection of planning, urban agriculture, and food justice: A review of the literature. *Journal of the American Planning Association, 83*(3), 277–295. https://doi.org/10.1080/01944363.2017.1322914

Howell, E., Harris, L. E., & Popkin, S. J. (2005). The health status of HOPE VI public housing residents. *Journal of Health Care for the Poor and Underserved, 16*(2), 273–285. https://doi.org/10.1353/hpu.2005.0036

Howerton, G., & Trauger, A. (2017). "Oh honey, don't you know?" The social construction of food access in a food desert. *ACME: An International Journal for Critical Geographies, 16*(4), 740–760.

Joint Center. (2012). *Place matters for health in Baltimore: Ensuring opportunities for good health for all* (A Report on Health Inequities in Baltimore, MD). Joint Center for Political & Economic Studies.

Keene, D. E., & Geronimus, A. T. (2011). "Weathering" HOPE VI: The importance of evaluating the population health impact of public housing demolition and displacement. *Journal of Urban Health, 88*(3), 417–435. https://doi.org/10.1007/s11524-011-9582-5

Kestens, Y., Lebel, A., Daniel, M., Thériault, M., & Pampalon, R. (2010). Using experienced activity spaces to measure foodscape exposure. *Health & Place, 16*(6), 1094–1103. https://doi.org/10.1016/j.healthplace.2010.06.016

Kwan, M.-P. (2009). From place-based to people-based exposure measures. *Social Science & Medicine, 69*(9), 1311–1313. https://doi.org/10.1016/j.socscimed.2009.07.013

LaDuke, W. (2019). *Indigenous food sovereignty in the United States: Restoring cultural knowledge, protecting environments, and regaining health* (D. A. Mihesuah, & E. Hoover, Eds.; Illustrated ed.). University of Oklahoma Press.

Lagisetty, P., Flamm, L., Rak, S., Landgraf, J., Heisler, M., & Forman, J. (2017). A multi-stakeholder evaluation of the Baltimore City virtual supermarket program. *BMC Public Health, 17*(1), 837. https://doi.org/10.1186/s12889-017-4864-9

Leung, M. M., Agaronov, A., Entwistle, T., Harry, L., Sharkey-Buckley, J., & Freudenberg, N. (2017). Voices through cameras: Using photovoice to explore food justice issues with minority youth in East Harlem. *New York. Health Promotion Practice, 18*(2), 211–220. https://doi.org/10.1177/1524839916678404

Levkoe, C. Z., Hammelman, C., Reynolds, K., Brown, X., Chappell, M. J., Salvador, R., & Wheeler, B. (2020). Scholar-activist perspectives on radical food geography: Collaborating through food justice and food sovereignty praxis. *Human Geography, 13*(3), 293–304. https://doi.org/10.1177/1942778620962036

Lipperman-Kreda, S., Mair, C., Grube, J. W., Friend, K. B., Jackson, P., & Watson, D. (2014). Density and proximity of tobacco outlets to homes and schools: Relations with youth cigarette smoking. *Prevention Science: The Official Journal of the Society for Prevention Research, 15*(5), 738–744. https://doi.org/10.1007/s11121-013-0442-2

Lynch, J., & Smith, G. D. (2005). A life course approach to chronic disease epidemiology. *Annual Review of Public Health, 26*(1), 1–35. https://doi.org/10.1146/annurev.publhealth.26.021304.144505

Manjarrez, C. A., Popkin, S. J., & Guernsey, E. (2007). *Poor health: Adding insult to injury for HOPE VI Families: (725572011-001)* (Data set). American Psychological Association. https://doi.org/10.1037/e725572011-001

McDowell, L. (1992). Doing gender: Feminism, feminists and research methods in human geography. *Transactions of the Institute of British Geographers, 17*(4), 399. https://doi.org/10.2307/622707

Morales-Campos, D. Y., Parra-Medina, D., & Esparza, L. A. (2015). Picture this!: Using participatory photo mapping with hispanic girls. *Family & Community Health, 38*(1), 44–54. https://doi.org/10.1097/FCH.0000000000000059

National Nurses United. (2019). *Burdening Baltimore: How Johns Hopkins Hospital and other not-for-profit hospitals, colleges, and universities fail to pay their fair share.* https://act.nationalnursesunited.org/page/-/files/graphics/0919_JHH_BurdeningBaltimore_Report-op.pdf

Perchoux, C., Chaix, B., Cummins, S., & Kestens, Y. (2013). Conceptualization and measurement of environmental exposure in epidemiology: Accounting for activity space related to daily mobility. *Health & Place, 21*, 86–93. https://doi.org/10.1016/j.healthplace.2013.01.005

Petteway, R., & Adams, L. (2011). Project food for thought: Youth perspectives on a Baltimore food desert. *139th meeting of the American Public Health Association.* American Public Health Association, Washington, DC.

Petteway, R., & Ames, A. (2011). *The 2011 Baltimore city neighborhood health profiles.* Baltimore City Health Department, Office of Epidemiology and Planning.

Petteway, R. J., Sheikhattari, P., & Wagner, F. (2019). Toward an intergenerational model for tobacco-focused CBPR: Integrating youth perspectives via photovoice. *Health Promotion Practice, 20*(1), 67–77. https://doi.org/10.1177/1524839918759526

Pies, C., & Parathasarthy, P. (2008). *Photovoice: Giving local health departments a new perspective on community health issues.* Contra Costa County Health Services, Public Health Division.

Pietila, A. (2010). *Not in my neighborhood: How bigotry shaped a great American city* (Illus. ed.). Ivan R. Dee.

Power, G. (1983). Apartheid Baltimore style: The residential segregation ordinances of 1910–1913. *Maryland Law Review, 42*(2), 43.

Ramírez, M. M. (2015). The elusive inclusive: Black food geographies and racialized food spaces. *Antipode, 47*(3), 748–769. https://doi.org/10.1111/anti.12131

Raskind, I. G., Kegler, M. C., Girard, A. W., Dunlop, A. L., & Kramer, M. R. (2020). An activity space approach to understanding how food access is associated with dietary intake and BMI among urban, low-income African American women. *Health & Place, 66*, 102458. https://doi.org/10.1016/j.healthplace.2020.102458

Reese, A. M. (2019). *Black food geographies: Race, self-reliance, and food access in Washington, DC* (Illus. ed.). University of North Carolina Press.

Richardson, D. M., & Nuru-Jeter, A. M. (2012). Neighborhood contexts experienced by young Mexican-American women: Enhancing our understanding of risk for early childbearing. *Journal of Urban Health, 89*(1), 59–73. https://doi.org/10.1007/s11524-011-9627-9

Richman, T. (2019). Activists rally for Hopkins, other nonprofits to 'pay their fair share' as Baltimore council reviews tax deal. *Baltimoresun.Com.* https://www.baltimoresun.com/politics/bs-md-pol-pilot-agreement-rally-20191219-i4fpxxzuhbg27kytw6jz6f6eja-story.html

Roche Jr., W. (2002). Housing plan found to do "more harm than good." *Baltimore Sun.* https://www.baltimoresun.com/news/bs-xpm-2002-06-28-0206280066-story.html

Sadler, R. C., Clark, A. F., Wilk, P., O'Connor, C., & Gilliland, J. A. (2016a). Using GPS and activity tracking to reveal the influence of adolescents' food environment exposure on junk food purchasing. *Canadian Journal of Public Health, 107*(1), eS14–eS20. https://doi.org/10.17269/CJPH.107.5346

Sadler, R. C., Gilliland, J. A., & Arku, G. (2016b). Theoretical issues in the 'food desert' debate and ways forward. *GeoJournal, 81*(3), 443–455. https://doi.org/10.1007/s10708-015-9634-6

Santo, R., Yong, R., & Palmer, A. (2014). Collaboration meets opportunity: The Baltimore food policy initiative. *Journal of Agriculture, Food Systems, and Community Development, 4*(3), 193–208–193–208. https://doi.org/10.5304/jafscd.2014.043.012

Song, H.-J., Gittelsohn, J., Kim, M., Suratkar, S., Sharma, S., & Anliker, J. (2009). A corner store intervention in a low-income urban community is associated with increased availability and sales of some healthy foods. *Public Health Nutrition, 12*(11), 2060–2067. https://doi.org/10.1017/S1368980009005242

Strack, R. W., Magill, C., & McDonagh, K. (2004). Engaging youth through photovoice. *Health Promotion Practice, 5*(1), 49–58. https://doi.org/10.1177/1524839903258015

Taylor, D. E., & Ard, K. J. (2015). Research article: Food availability and the food desert frame in Detroit: An overview of the city's food system. *Environmental Practice, 17*(2), 102–133. https://doi.org/10.1017/S1466046614000544

Trauger, A. (2017). *We want land to live: Making political space for food sovereignty.* University of Georgia Press.

Villanueva, K., Giles-Corti, B., Bulsara, M., McCormack, G. R., Timperio, A., Middleton, N., Beesley, B., & Trapp, G. (2012). How far do children travel from their homes? Exploring children's activity spaces in their neighborhood. *Health & Place, 18*(2), 263–273. https://doi.org/10.1016/j.healthplace.2011.09.019

Wang, C. C. (1999). Photovoice: A participatory action research strategy applied to women's health. *Journal of Women's Health, 8*(2), 185–192.

Wang, C. C. (2006). Youth participation in photovoice as a strategy for community change. *Journal of Community Practice, 14*(1–2), 147–161. https://doi.org/10.1300/J125v14n01_09

Wang, C., & Burris, M. A. (1997). Photovoice: Concept, methodology, and use for participatory needs assessment. *Health Education & Behavior: The Official Publication of the Society for Public Health Education, 24*(3), 369–387. https://doi.org/10.1177/109019819702400309

Wensel, C. R., Trude, A. C. B., Poirier, L., Alghamdi, R., Trujillo, A., Anderson Steeves, E., Paige, D., & Gittelsohn, J. (2019). B'more healthy corner stores for moms and kids: Identifying optimal behavioral economic strategies to increase WIC redemptions in small urban corner stores. *International Journal of Environmental Research and Public Health, 16*(1). https://doi.org/10.3390/ijerph16010064

Wethington, E., & Johnson-Askew, W. L. (2009). Contributions of the life course perspective to research on food decision making. *Annals of Behavioral Medicine, 38*(suppl_1), s74–s80. https://doi.org/10.1007/s12160-009-9123-6

Widener, M. J., Minaker, L. M., Reid, J. L., Patterson, Z., Ahmadi, T. K., & Hammond, D. (2018). Activity space-based measures of the food environment and their relationships to food purchasing behaviours for young urban adults in Canada. *Public Health Nutrition, 21*(11), 2103–2116. https://doi.org/10.1017/S1368980018000435

Zenk, S. N., Schulz, A. J., Matthews, S. A., Odoms-Young, A., Wilbur, J., Wegrzyn, L., Gibbs, K., Braunschweig, C., & Stokes, C. (2011). Activity space environment and dietary and physical activity behaviors: A pilot study. *Health & Place, 17*(5), 1150–1161. https://doi.org/10.1016/j.healthplace.2011.05.001

Chapter 3
Placescapes + Public Housing: Toward a Critical Understanding of "Place" + "Placemaking" in Place-Based Health and Housing Strategies

(Mis)Representation

Interrogating Notions of "Place" and Health in the Projects

Place-based strategies are increasingly being explored as options to improve health, education, and general life opportunities among poor and marginalized communities in the United States (HCZ, n.d.; Maryland DHMH, n.d.; NCHE, n.d.; TCE, n.d.; Whitehurst & Croft, 2010). This is especially true for residents of public housing, as many of the prominent place-based strategies to date have been federal initiatives involving the Department of Housing and Urban Development (HUD), e.g., HOPE VI, Promise Zones, Choice Neighborhoods, and Sustainable Communities (HUD, 2013a, b, d, e, 2015c). Health status among public housing residents is generally much worse compared to the general population, and public housing communities tend to be located within areas having particularly noxious built, social, and economic environments (Buron et al., 2002; Digenis-Bury et al., 2008; Fertig & Reingold, 2007; Harris & Kaye, 2004; Howell et al., 2005; Keene & Geronimus, 2011; Manjarrez et al., 2007; Popkin, 2004; Popkin et al., 2004; Ruel et al., 2010). This leaves public housing residents, especially, in need of interventions and policies aimed at improving health opportunities. Moreover, from a public health prevention and lifecourse perspective, place-based strategies for health promotion involving public housing make intuitive sense—they're fixed, densely populated communities, and nearly 40% of residents are under the age of 18 (HUD, 2013c).

The more recent iterations of place-based strategies have been increasingly comprehensive, moving beyond simple considerations of housing quality and aiming to coherently link affordable housing opportunities with health, education, and transportation opportunities (HUD, 2013a, d, e). On the other hand, these newer and evolving strategies are being implemented in only a few selected cities and regions, and many jurisdictions are precluded from programmatic support due to population

© Springer Nature Switzerland AG 2022
R. J. Petteway, *Representation, Re-Presentation, and Resistance*, Global Perspectives on Health Geography, https://doi.org/10.1007/978-3-031-06141-7_3

size requirements. Thus, the overwhelming majority of project-based public housing continues to operate without the benefit not only of new money streams, but perhaps more importantly, new idea streams.

Moreover, many of the more prominent and large-scale place-based efforts to date have failed to engage notions of participation, power, and (dis)possession in their attempts to re-imagine, re-design, and revitalize "place" (Chaskin, 2013; Clampet-Lundquist, 2004a, b; Keene & Geronimus, 2011; NHLP, 2002; Slater, 2013). A consequence has been the continued problematizing and dislocation of *people* and the re-appropriation and re-constitution of their *place* as the solution. Not only have these efforts not led to many significant or consistent improvements in public housing resident well-being (Acevedo-Garcia, 2004; Clampet-Lundquist, 2004a, b; Fauth, 2008; Goetz, 2010; Goetz & Chapple, 2010; Harris & Kaye, 2004; Howell et al., 2005; Jones & Paulson, 2011; Keene & Geronimus, 2011; Leventhal & Brooks-Gunn, 2003; Levy & Woolley, 2007; Ludwig, 2011; Manjarrez et al., 2007; Popkin, 2004; Popkin et al., 2004), the manner in which they have been developed and implemented has systematically precluded resident agency—habitually circumventing critical examination of the underlying social, economic, and political structures that necessitate place-based strategies in the first place (Chaskin, 2013; Goetz, 2013a, b; Keene & Geronimus, 2011; NHLP, 2002; Slater, 2013; Keane & Geronimus, 2011; Chaskin, 2013; Goetz, 2013a, b). Accordingly, it is of critical importance to understand that many public housing "places" are pre-made—residents are commonly dispossessed (displaced) of one place and then dispersed (re-placed) into new locations where they have little control and few social or political connections. Thus, a requisite to understanding residents' lived experience with place, and consequent health effects, is explicating the mechanisms that either facilitate or limit their ability and power to participate in and influence placemaking, as both a social and material process.

Unfortunately, there is a paucity of public health research focused on residents of public housing. This lack of research not only limits our knowledge of the health status of public housing residents and how residing in public housing might influence health, but it also restricts our ability to understand how public housing fits into residents' larger geographic, social, and economic landscapes. For example, project-based public housing residents must move between many places outside their community to meet their daily needs and complete daily functions. While this is generally true for residents of single-family homes, it is especially important for public housing residents given their already vulnerable position and the added pressure to maintain their housing subsidy (Manjarrez et al., 2007). Moreover, there is often a spatial mismatch between where public housing residents live and the amenities and opportunities they need to sustain themselves (e.g., education and employment opportunities, grocery stores, healthcare facilities, pharmacies, and post offices). These non-residential places constitute a significant portion of their daily health-related opportunities and exposures. The extent to which the housing location and its external connectivity influence residents' ability to meet their daily needs is instrumental to their overall well-being. Thus, understanding how public housing fits into the larger spatial, social, economic, and political landscape of residents' lived place beyond the housing community boundaries—e.g., their "geographies of opportunity" (de Souza Briggs, 2005; Galster & Killen, 1995; Osypuk & Acevedo-Garcia, 2010; Petteway, 2021; Squires & Kubrin, 2005)—is critical to evaluating and improving place-based strategies involving public housing.

A central challenge is how to account for the daily space-time patterns of individuals and populations (who are simply residing in a particular location) to best design, implement, and evaluate place-based strategies that are sensitive to peoples' lived realities of place. Being able to do so would improve our ability to optimize spatial and social configurations of health assets and opportunities, while simultaneously minimizing negative place-based health exposures. In the context of place-based strategies involving public housing, this means understanding how the spatial location of the housing community fits within the daily *places* of its residents—where are the jobs, schools, parks, fresh food vendors, social hubs, pharmacies, health care providers, transportation hubs, and so on; what are the temporal, spatial, and social connections (or divisions) between these places; and where are the negative health exposures situated within these space-time configurations (e.g., at work, the walk to school, near the park). Thus, place-based strategies involving public housing would do well to critically assess and be responsive to the very person-centered spatiotemporal activity patterns of affected residents. This perspective would facilitate a more comprehensive understanding of health in public housing, and how to improve it, as well as ensure that place-based thinking maintains an appreciation for the individual and collective lived realities of residents—that is, a people-centered focus within place-based strategies.

The field of place-health research has grown rapidly over the last two decades (Arcaya et al., 2016; Diez Roux & Mair, 2010; Ellen et al., 2001; Pickett & Pearl, 2001; Riva et al., 2007; Sampson et al., 2002; Santos et al., 2007), and it is well-suited to help understand health in the context of public housing. However, as discussed in Chap. 1, major conceptual and methodological challenges remain in defining "place," characterizing place contexts, and measuring place—all of which have implications for place-health research, public health practice, and the design and implementation of place-based strategies. In the context of US public housing, projects are indeed clearly defined and fixed locales, and while they are undoubtedly in a particular census tract or set of tracts, residents' lives are not bound within them. The housing project is a singular node in their "spatially polygamous" and multi-nodal lives (Cummins et al., 2007; Matthews, 2011). Exposures and opportunities encountered within the project or immediate surrounding "neighborhood" are only a fraction of all those influencing their health status. If we want to better understand health of/among public housing residents, we need to better understand and account for how their place of residence is connected to—and shaped (even determined) by—their broader spatial, social, and political contexts. Of particular need is work capable of revealing: (1) spatially and temporally specific configurations of place-based exposures and opportunities, (2) perspectives and influences of place across generations and over the lifecourse, and (3) opportunities for action to address place exposures that adversely affect community health.

Accordingly, the goal of the work presented here was to develop and field-test a place-health framework that: (1) accounts for the multi-nodal nature of "place" and its contingent spatial, temporal, and social inter-nodal connections/divisions; (2) elucidates potential intergenerational and life-stage differences in place experiences/perceptions; and (3) explicitly engages the sociopolitical mechanisms that make, unmake, and remake place over time, i.e., social and material placemaking processes—shaping spatiotemporal patterns and sociospatial arrangements of place exposures and opportunities.

A framework for the *placescape* approach was developed drawing from critical theory, place-health, social epidemiology, participatory research, geography, and sociology literatures. This framework was then applied to an intergenerational community-based participatory research (CBPR) study of place, embodiment, and health. In the following sections, I first introduce the placescape framework and summarize its conceptual roots and core tenets. I then present the process and findings from an application of the framework. I close with a discussion of implications of this work for intergenerational place-health research/practice and place-based health and housing strategies going forward.

Re-Presentation, Pt. I

Toward a **Placescape** *Approach: Conceptual Roots and Core Tenets*

Background

Current dominant approaches to place-health research are quite limited in their ability to account for peoples' lived spatial realities, thus affording only a partial rendition of relevant place-based health exposures, health opportunities, and their related sociopolitical determinants. Accordingly, there is a need for enhanced approaches to understanding and studying place. Here, the *Placescape* is introduced as both an analytical framework and a conceptual orientation for understanding place that can more adequately capture the lived reality of place and its social, economic, and political determinants, and accordingly better inform the development, implementation, and evaluation of place-based health strategies—especially those involving public housing. The Placescape framework draws from place-health literature, as well as from ecosocial theory (Krieger, 1994, 2001), geography, and critical theory literatures. Specifically, Table 3.1 shows the concepts that form the foundation for the Placescape framework. The first six concepts, *"relational" place* through *"space-time constraints,"* deal primarily with considerations of how "place" is conceived, defined, measured, and operationalized. Taken together, these concepts outline the value of an approach to "place" that is spatially and temporally dynamic and bound not by imaginary administrative lines (Petteway, 2017, 2018), but by the daily movements and spatial behavior patterns of residents. The remaining concepts,

Table 3.1 Conceptual foundations for the placescape framework in public health and public housing

Concept/ construct	Author(s)	Theoretical home/ field(s)	Summary
Relational place	Cummins et al. (2007)	Place-health research; public health; health geography	• "Place" is not a singular and static spatial location, but rather a network of locations that vary by person over time and over the lifecourse • Connections between these places are best understood in terms of "social-relational distance," as opposed to simply physical distance • Spatial and territorial divisions do not necessarily coincide with administrative boundaries, and are "imbued with social power relations and cultural meaning" (p. 1827) and are experienced viscerally and corporally by residents moving through multiple places during their day-to-day activities and over their lifecourse • Places change over time and are shaped by processes at both local and non-local levels
Opportunity structures	Macintyre et al. (2002)	Place-health research; public health; health geography	• Refers to "socially constructed and socially patterned features of the physical and social environ-ment which may promote or damage health either directly, or indirectly through the possibilities they provide for people to live healthy lives" (p.132) • These structures have particular spatial arrangements within a local geographic context and these arrangements are shaped by both local and non-local social, economic, and political processes, policies, and practices. • Opportunity structures are actively made, and accordingly can be either preserved as is or modified, with consequent effects on residents' lived place

(continued)

Table 3.1 (continued)

Concept/construct	Author(s)	Theoretical home/field(s)	Summary
Needs-driven place	Macintyre et al. (2002)	Place-health research; public health; health geography	• People have a set of basic human needs which need to be met to live healthy lives • A resident's needs-driven place is in part governed by the opportunity structures they encounter and experience within the context of their daily lives—Which may necessitate, even dictate, that they meet their needs across spatially distant and socially disparate places
Spatial polygamy	Matthews (2011) and Matthews and Yang (2013)	Place-health research; public health; health geography	• Rooted in the idea that people are not "loyal" to a singular place, but have an affinity for and meaningful connections to multiple places • People have core anchor points, or "nodes," in their daily lives, and residence is just one • The spatial distributions of these place nodes and their inter-nodal connections do not readily coincide with traditional spatial bounds used in place-health research
Activity space	Golledge and Stimson (1996), Jones and Pebley (2014), Kwan (2009), Perchoux et al. (2013), and Wong and Shaw (2011)	Human geography; space-time geography; transportation; place-health research	• Refers to the geographic spaces people travel to and within during their day-to-day activities • Generally composed of a set of both "fixed" and "flexible" places, structured around core "anchors" or "nodes." • Has both spatial (geographic locations and routes to/from places encountered) and temporal (e.g., frequency, regularity, duration, sequencing, and timing of place encounters) emphases • A way to assess and account for peoples' mobility patterns and spatial behaviors, and thus their spatially and temporally specific health-related opportunities and exposures as experienced through their daily lives • The spatial characteristics of activity spaces do not readily coincide with traditional spatial bounds used in place-health research, e.g., census tracts, ZIP codes

(continued)

Table 3.1 (continued)

Concept/ construct	Author(s)	Theoretical home/ field(s)	Summary

Space-time constraints	Hägerstrand (1970) and Kwan (2000)	Human geography; space-time geography; feminist geography	• Generally, a combination of limitations/demands that determine the spatial and temporal character of an individual's daily activities
			Three categories:
			  • *Capability constraints*: Limits to space-time activities imposed by physiological capacities (e.g., needing to sleep and eat) or distance
			  • *Coupling constraints*: Limits to space-time activities imposed by the need of other people, resources, tools, etc. to complete the activity, e.g., meetings, appointments.
			  • *Authority constraints*: Limits to space-time activities imposed by formal institutions (in the broad sense), e.g., work hours, business hours, school hours; also, limits imposed more informally (e.g., by individuals) that restrict access to space for certain people
			• In the context of place-health relationships and housing, space-time constraints help contextualize resident spatial behaviors in relation/response to, for example, existing opportunity structures and their daily needs or experiences of spatialized social exclusion

Riskscape	Morello-Frosch and Lopez (2006) and Morello-Frosch and Shenassa (2006)	Public health; environmental justice; environmental health research	• Describes the myriad of environmental health exposures that tend to overlap (temporally and spatially) within low-income and segregated communities of color • The spatial distribution and concentration of these exposures are shaped by current and historical social, economic, and political practices and policies that devalue and disproportionately burden the community environments of socially disadvantaged populations

(continued)

Table 3.1 (continued)

Concept/ construct	Author(s)	Theoretical home/ field(s)	Summary
Pathways of embodiment	Krieger (2005) and Krieger (2001)	Ecosocial theory; public health; social epidemiology	• Within ecosocial theory, *embodiment* describes the process through which the outside material and social world becomes biologically incorporated • *Pathways of embodiment* are the avenues through which social inequality, power imbalances, and resource inequities shape and constrain life opportunities with consequent effects on our physiologic functioning
Agency and accountability	Krieger (1994, 2001)	Ecosocial theory; public health; social epidemiology	• Considers who is responsible for shaping and maintaining the societal arrangements of power, resources, and opportunity (i.e., the pathways of embodiment) and. Thus. Accountable for consequent health inequity • In the context of "place," this construct challenges us to think critically about how historic and current power relations have patterned distributions of resources and opportunities both socially and spatially • Challenges us to think explicitly about responsibility and culpability, asking who and what is responsible in shaping, maintaining, and/or mitigating inequities, including consider-ations of agency and power within knowledge production regarding such inequities

(continued)

Table 3.1 (continued)

Concept/ construct	Author(s)	Theoretical home/ field(s)	Summary
Accumulation by dispossession	Harvey (2004, 2006)	Critical theory; critical geography	• Describes how goods/assets are systematically transferred from the masses (i.e., public) to the upper class (i.e., private or class-privileged), driving a process of *uneven geographical development* • These "transfers" are necessitated by crises of capital over-accumulation—new spaces of development are needed for continued value gains and growth • The major modalities of accumulation by dispossession include privatization, financialization, the management and manipulation of crises, and state redistributions • In the context of place and public housing, examples might include strategies involving mixed-income housing developments, the deliberate deterioration and devaluation of properties, demolition of public housing and subsequent use of the space for private benefit, demolition of public housing without 1-to-1 replacement, and displacement of residents under the auspices of community development or "revitalization."

(continued)

Table 3.1 (continued)

Concept/ construct	Author(s)	Theoretical home/ field(s)	Summary
Circuits and consequences of dispossession	Fine and Ruglis (2009) and Ruglis (2011)	Critical theory; education	• Describes how neoliberal processes and mechanisms of accumulation by dispossession function to de-value and deprive socially disadvantaged youth of equitable education resources and opportunities • Students are dispossessed of their right to a social good—educa-tion—the loss of which is "offset" by private gains • A "circuit" of dispossession is formed by various state-sanc-tioned education (and non-educa-tion) policies/practices. A "consequence" of this disposses-sion is compromised health • In relation to housing and health, it extends the notion of accumula-tion by dispossession by more explicitly articulating *embodied* consequences, not just capital/ spatial consequences, of dispossession

(continued)

Table 3.1 (continued)

Concept/ construct	Author(s)	Theoretical home/ field(s)	Summary
Biopower	Foucault and Senellart (2008)	Critical theory; biopolitics	• Refers to the myriad of technologies of power that center on managing, regulating, and subjugating physical bodies (people). In the context of place, health, and public housing in the United States, these technologies include programs, policies, and practices like Urban Renewal, HOPE VI, Choice Neighborhoods, Community Development Block Grants, Low-Income Housing Tax Credit, New Markets Tax Credit, and Opportunity Zones via the 2017 Tax Cuts and Jobs Act • This ordering and manipulation of populations of bodies, accompanied by an array of techniques (e.g., statistics, laws) and discursive practices (e.g., science, knowledge production), is used to (re)produce and justify particular social and political arrangements (e.g., of goods, property, opportunity) • Technologies of biopower are thus a primary mechanism through which bodies are socially and spatially organized and controlled as relevant to health risks/ exposures, e.g., vis a vis structuring "geographies of opportunity" (de Souza Briggs, 2005; Galster & Sharkey, 2017; Powell & Bullard, 2007) and pathways of embodiment (Krieger, 2005)

(continued)

Table 3.1 (continued)

Concept/ construct	Author(s)	Theoretical home/ field(s)	Summary
Primacy of racialization	Delgado et al. (2017) and Ford and Airhihenbuwa (2010)	Critical Race Theory; Public Health Critical Race Praxis; social epidemiology	• Refers to the "fundamental contribution of racial stratification to societal problems" (Ford & Airhihenbuwa, 2010, p. 1394) • In the context of place-based health risks, exposures, and opportunities, racialization can be seen as a core process underlying inequitable built, social, and economic environmental conditions, e.g. vis a vis (de) valuations of a place's residents • Racialization also shapes broader narratives, representations, and reputations of place, i.e., racialization of space and creation/propagation of spatial stigma hall (Cairns, 2018; Graham et al., 2016; Halliday et al., 2020; Jewell, 2018; Tran et al., 2020)
Structural determinism	Delgado et al. (2017) and Ford and Airhihenbuwa (2010)	Critical Race Theory; Public Health Critical Race Praxis; social epidemiology	• Refers to the "fundamental role of macro-level forces in driving and sustaining inequities across time and contexts" (Ford & Airhihenbuwa, 2010, p. 1394) • In the context of place-health relationships as connected to housing, structural determinism is observed, for example, in the practices and processes of redlining, racist zoning ordinances, racially restrictive covenants, urban renewal, and gentrification • Serves as a foundational principle for historicizing and contextualizing the spatial sorting of people in relation to health and housing opportunity, and for coloring discourse of biopower, accumulation, and dispossession as related to place

(continued)

Table 3.1 (continued)

Concept/ construct	Author(s)	Theoretical home/ field(s)	Summary
White possessive logics	Moreton-Robinson (2015)	Critical theory; decolonial theory; indigenous studies	• Describes, "a mode of rationalization… that is underpinned by an excessive desire to invest in reproducing and reaffirming the nation-state's ownership, control, and domination" (p.xii); here, also extending into ownership/control of data regarding and representations of "place." • Involves multiple and varying "significations" of White ownership/control as physically marked within landscapes/ cityscapes and/or socially embedded within sociopolitical spatial relationships that mirror the geographic contours and consequences of racialization • Serves as a foundational logic for continual spatial dispossession, (re)possession, and erasure observed in the context of urban land use, community development, and housing practices

Summary of core theoretical and conceptual roots that inform/animate the Placescape framework as outlined in this chapter. The intention is not to be exhaustive, but rather to synthesize some key concepts in order to outline a cursory framework for more critically engaging/accounting for notions of power and/in placemaking as germane to place-health research, vis a vis the (re)production of social and material place-health contexts—and sociospatial distributions and patterns of health risks/opportunities therein. The "Summary" column articulates some of this, here as particularly relevant to place and health in the context of US public housing

"riskscape" through *"White possessive,"* deal primarily with considerations of power, agency, and accountability within historic and current policies, practices, and processes that fundamentally shape the social, economic, and environmental character of residents' places, and the spatial patterns and distributions of health opportunities and risks within, i.e., placemaking mechanisms. These concepts highlight the critical import of taking an approach to place that explicitly acknowledges and engages the manner in which place is actively (and continuously) made, unmade, and remade, and is thus attuned to the structured yet malleable nature of residents' place-based health exposures and opportunities.

The concepts outlined in Table 3.1 form a rich theoretical foundation for developing the Placescape as a cursory analytic framework through which to view and appraise "place" and the manner and processes through which it is made, unmade, and remade over time. In the context of US public housing, these placemaking processes include not only public housing strategies and policies (e.g., Choice Neighborhoods, Sustainable Communities, Section 8) (HUD, 2013a, b, e, 2015e), but also mechanisms like the Community Development Block Grant, or CDBG (HUD, 2015a), the low-income housing tax credit, or LIHTC (HUD, 2015f), Home Investment Partnership Program, or HOME (HUD, 2015d, h), and the Community Reinvestment Act, or CRA (FDIC, 2015). These placemaking mechanisms, representing both "pathways of embodiment" and technologies of "biopower," are themselves influenced by larger social, economic, and political realities (Bloom et al., 2015; Goetz, 2013a), and ultimately shape residents' lived and embodied experiences of place—their individual and collective *placescapes*. The extent to which placemaking processes are inclusive, equitable, and attentive to residents' basic needs and well-being is thus a key determinant of the overall spatial structure of health opportunity and risk in relation to public housing, and this structure, in turn, shapes residents' daily mobility patterns and spatial movement, thus the centrality of needs-driven place, "relational" place, "opportunity structures," "spatial polygamy," "activity space," and "riskscape." This larger placescape framework challenges us to actively engage "place" in a manner that accounts for the historically dynamic, socially malleable, and economically and politically contingent nature of how it is (materially), how it is perceived/narrated, and how it came to be—thus the import of "agency and accountability," "accumulation by dispossession," "circuits and consequences of dispossession," "primacy of racialization," "structural determinism," and "White possessive." One potential way to elucidate how these processes unfold, and how they are experienced to influence health opportunity and risk, is through engaging residents in processes of participatory research. Drawing from the concepts in Table 3.1, applying a placescape orientation to place-health research entails a few core tenets, summarized in Table 3.2.

Figure 3.1 is a visual schematic of the Placescape framework in the context of health in public housing. The large red circle represents a *circuit of dispossession*. The blue circles represent expressions/technologies of *biopower*. The green circle represents the *placescape* paradigm (described in Table 3.2 above) as experienced by residents, i.e., their spatially and temporally specific experiences, and life-stage contingent sociomaterial contexts and perceptions of place. Note that each of the circles representing *biopower* and the *placescape* are elements of the larger circuit of dispossession, but also function independently as circuits themselves (as represented by the red circle outlining around each). The blue arrow moving from the *placescape* to *health* represents the link through which the larger *circuit* of dispossession becomes a *consequence* of dispossession (here, health). For example, housing and community development policy, as technologies of biopower, form and interact within a circuit of dispossession, making and remaking residents' lived placescapes (e.g., via spatial ordering and sorting of people; modification of social networks, built and natural environments, economic and political contexts; displacement and spatial dispossession). This manifestation of biopower, expressed at the population level, is encountered and experienced via residents' lived *placescape*; and this *placescape* experience is embodied by residents *individually* (i.e., spatiotemporal and sociospatial patterns of place-based exposures). Thus, the blue coloring of the arrow between the *placescape* and *health* indicates the link between technologies of biopower at the population level (e.g., public housing policy) and the physiological embodiment of place at the individual level, as mediated by

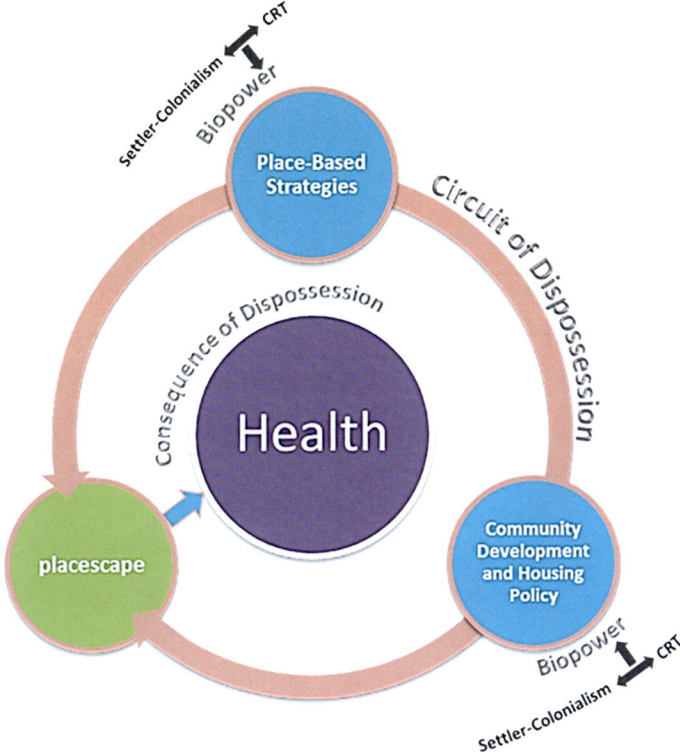

Fig. 3.1 A placescape framework for place and health in public housing. Cursory conceptual and analytic Placescape framework for understanding various aspects of power as relevant to the production of social and material contexts of consequence for place, health, and public housing. Drawing from concepts outlined in Table 3.1, this figure visually illustrates conceptualized relationships between core placemaking mechanisms pertinent to place and health in the context of public housing

residents' lived placescape. Importantly, in the context of population health inequities as shaped by place, biopower is by no means universally/uniformly exercised. Here, it reflects the United States' settler-colonial and White supremacist roots. As such, no conception of "place" is possible without an explicit account for what Moreton-Robinson (2015) refers to as the "White possessive," and what critical race scholars refer to as the "primacy of racialization" and "structural determinism" (Delgado et al., 2017; Ford & Airhihenbuwa, 2010). Space is unmistakably racialized, and race is unmistakably spatialized (de Souza Briggs, 2005; Edwards & Thomson, 2010; Galster & Sharkey, 2017; Kent-Stoll, 2020; Neely & Samura, 2011; Powell & Bullard, 2007; Powell & Cardwell, 2013; Shabazz, 2015)—especially in the context of public housing. Moreover, these processes remain rooted in spatial dispossession, (re)possession, and epistemic erasure. These concepts thus inform how biopower functions—i.e., as an explicitly racialized and settler-colonial expression of spatial control/dominance—in the contexts of place-based strategies and community development/housing policy.

Table 3.2 The placescape: core operational tenets for place-health research

Placescape tenet (PT)	Summary
PT1: Needs & Opportunities	Each resident has a unique set of needs; thus, any notion of "place" must account for similarities and differences between residents' needs-driven configuration of places and the place-based opportunity structures that shape how and where such needs are met
PT2: Mobility & Bounds	Residents viewed as actors with variant mobility patterns over time, and these patterns do not necessarily coincide with administrative bounds. Mobility is both enabled and constrained by social, political, and economic factors/actions that influence access to, control over, and acceptance within each "place."
PT3: Multinodal place	"Place" is best seen as a particular configuration of "nodes" with connections, divisions, and restrictions that constitutes a person's lived spatiotemporal place network
PT4: Power in place (making)	Place configurations are consequences of historic and pervading power relations. Within these power relations, *place is both made and re-made, both consumed and produced, and both includes and excludes* (a) Distributions and degree of benefit and harm from each process are inextricably linked to distributions of power and participation underlying each process (b) Considerations of who is responsible for shaping and maintaining societal arrangements of power and resources—particularly as related to space (e.g., access, (dis)possession, and (de)valuation), and the sociopolitical and economic mechanisms underlying the spatial sorting of opportunity—are critical
PT5: Lifecourse in place	Place experiences and perceptions are not static or universal, but best seen as time-variant and generationally and life-stage contingent; as such, place effects on health are best viewed as the product of the space- and time-specific exposures (positive and negative) residents encounter in their daily lives and over their lifecourse, with attention to multigenerational effects
PT6: Agency in place (making)	Place-health research *is part of the placemaking process* (a) All entities/persons are actors with varying degrees of knowledge, expertise, and power whose expressions and manifestations are implicated in either the maintenance of or challenge to current conditions (b) As such, communities should be proactively engaged in the research process, as their embodied experiences, perspectives, and expertise can improve place-health research, research translation, and intervention efforts

Overview of six core *placescape tenets* as conceptualized for place-health research, drawing from the concepts outlined in Table 3.1. These tenets offer an orientation to place-health research that remains sensitive to various aspects of power as relevant to the (re)production of social and material contexts of "place," i.e., placemaking, which includes place-health knowledge production. They also center on the critical import of spatially dynamic and relational approaches to place-health research that can accommodate multiple and interrelating representations/articulations of peoples' place-health experiences

Re-Presentation, Pt. II

The Placescape in Practice: An Intergenerational Study of Place, Embodiment, and Health

Background

The People's Social Epi Project (PSEP) was developed and executed with an orientation anchored in *A People's Social Epidemiology* framework (Petteway et al., 2019b)—a multicomponent and tiered framework to guide social epidemiology research/practice to become more inclusive and equitable, improve knowledge translation, and facilitate timely, locally relevant action. The placescape approach was applied within PSEP to examine and demonstrate its utility in modeling *A People's Social Epi* within place-health research. The PSEP integrated social epidemiology and community-based participatory research (CBPR) in collaboration with parents and youth residing in public housing to further understand where and how place-based exposures that affect health and well-being are encountered, perceived, and experienced intergenerationally. This work sought to: (1) expand and make novel contributions to research on health in public housing; (2) improve conceptual and operational understandings of place through identifying the spatial, temporal, and social connections and divisions between the places of residents' daily activities; and (3) elucidate spatial, temporal, and perceptual differences between parent and youth place experiences. The specific aims for the work presented here were to:

1. Determine the spatial distribution of adult and youth daily places within 5 broad place-domains: *Home*, *Neighborhood*, *School/Work*, *Social/Leisure*, and *Transition.*
2. Characterize adult and youth perceptions of place-embodiment for their daily places.
3. Assess spatial differences of "place" and perceptual differences of place-embodiment between adults and youth.

The research was completed using participatory methods for the systematic documentation and assessment of place-based exposures and opportunities with two generations of public housing residents recruited from a predominantly Black public housing project located in Steubenville, OH—a small rustbelt city just outside Pittsburgh, PA. This housing community represented one of only a few remaining affordable housing options in the participants' city, and at the time of this study, residents were feeling particularly concerned about their current and future housing prospects, due to attendant development pressures (discussed below). This context made the current project especially timely and relevant.

All recruitment and project activities were completed with guidance from an adult resident research co-lead who was trained in human subjects research. One parent and at least one youth from each participating household were recruited as parent-child dyads. Youth were between ages 13 and 17 and had to be enrolled in school. Parents had to have some daily form of formal or informal employment or non-leisure activity (e.g., job, childcare, doing friends' hair, errands). A total of 8 adults and 10 youth participated in the project. The process and findings presented here are from the first iteration of the PSEP, for which complete data are available for four adults and seven youth.

Process + Methods

Participants were trained in key components of public health including core principles related to social epidemiology and health equity, and fundamental aspects of public health research and CBPR. All research methods were participatory and completed by the participants themselves. Research methods flowed sequentially and built upon each other as follows: (1) *Photovoice* (Catalani & Minkler, 2010; Wang & Burris, 1997); (2) *Activity Space Mapping* (see, for example, Browning & Soller, 2014; Chaix et al., 2012; Perchoux et al., 2013); (3) *X-Ray Mapping*; and (see Ruglis, 2011) (4) *Participatory GIS*. Adults and youth completed each method simultaneously but in separate all-adult and all-youth groups.

First, participants used photovoice (via cellphones) to identify, photo-document, and describe important daily places and specific exposures/opportunities within each place they perceived affected their health (positively or negatively). They took photos on their own time as they went about their daily lives. They completed analytic narratives for each photo using the *SHOWeD* photovoice inductive questions guide (see Wang, 1999), and then completed their own participatory qualitative analysis to identify core themes (detailed in Petteway, 2019).

Second, participants used a participatory *Activity Space Mapping* (ASM) method to geolocate and map their photovoice photos and identify any additional non-photographed places. Briefly here, they used large print-out maps to identify the locations of their *Photovoice* photos using stickers and markers. Participants then identified additional important places for which they had not taken photos. They then completed an ASM worksheet for each identified location to rate (e.g., how healthy/unhealthy) and provide estimates of time spent in each mapped place (detailed in Chap. 4).

Third, using a cognitive mapping method known as *X-Ray Mapping,* they created symbolic representations of place-embodiment reflecting how each of their mapped places affected their bodies and health. Briefly here, this involved using $8.5'' \times 11''$ worksheets with a basic body outline and color-coded stickers to indicate how their bodies were affected by their daily places (detailed in Chap. 5).

Finally, constituting participatory GIS, study participants integrated and digitally mapped their work via a web-based interactive and multimedia-enabled information and communication technology (ICT) platform, *Local Ground* (Van Wart et al., 2010). This platform allowed participants to easily create, enhance, print, and digitally share their place research maps with the broader community and city officials.

Participants' data elements (e.g., photos, activity space maps, X-Ray Maps) were assigned to a place-domain based on the data topic and location. For example, a photo of a participant's housing environment would be assigned to the "Home" domain, and a photo related to a participant's school/place of work would be assigned to the "School/Work" domain, and so on. Data reflecting their community built, social, food environments, etc. were assigned to the "Neighborhood" domain, except those data for which associated narratives indicated that a particular location was simply observed/passed on their route/way to another intended destination (e.g., "I walk by this building on the way to school"). In this case, data were assigned to the "Transition" domain *and* "Neighborhood" domain, but counted only in the "Transition" domain for the data presented here. Data related to leisure/social activities or related places were assigned to the "Leisure/Social" domain.

Findings

Youth took a total of 66 photos during photovoice, of which they selected 31 for inclusion in their participatory theming, coding, and ranking analysis (Petteway, 2019). However, sufficient information was available to geolocate, assign a place-domain category, and determine positive/negative participant place appraisals for 47 photos. Adults took 49 photos total, 20 of which they included in their participatory analysis. For adults, there was sufficient information for 31 photos for geolocation, place-domain categorization, and positive/negative determinations. Tables 3.3 and 3.4

Table 3.3 Youth photovoice results summary by place-domain

Youth Photovoice Results Summary			
Place-Domain	# of Photos	# Positive/Healthy/Good Place	# Negative/Unhealthy/Bad Place
Home	7	0 (0%)	7 (100%)
Neigborhood	23	6 (26%)	17 (84%)
School	6	1 (17%)	5 (83%)
Leisure/Social	6	6 (100%)	0 (0%)
Transition	5	1 (20%)	4 (80%)
Total	47	14 (30%)	33 (70%)

Summary of youth photovoice photos and place-health perceptions by place-domain. Youth took photos of specific places and characterized them as either "positive/health/good," "negative/unhealthy/bad," or both. Each photo was then assigned a place-domain

Table 3.4 Adult photovoice results summary by place-domain

Adult Photovoice Results Summary			
Place-Domain	# of Photos	# Positive/Healthy/Good Place	# Negative/Unhealthy/Bad Place
Home	10	2 (20%)	8 (80%)
Neigborhood	15	2 (13%)	13 (87%)
Work/Errand	0	0	0
Leisure/Social	4	4 (100%)	0 (0%)
Transition	2	0 (0%)	2 (100%)
Total	31	8 (26%)	23 (74%)

Summary of adult photovoice photos and place-health perceptions by place-domain. Adults took photos of specific places and characterized them as either "positive/health/good," "negative/unhealthy/bad," or both. Each photo was then assigned a place-domain

summarize youth and adult photovoice data across the 5 place-domains in regard to how they appraised their photo-places. Figures 3.2 and 3.3 show an example of how participants geolocated their photovoice photos (and narratives) using the Local Ground platform. The photos chosen for presentation in Figs. 3.2 and 3.3 were selected based on how prominent or important participants viewed their photovoice photos—they represent the photos that received the most participatory analysis codes within the thematic photo group voted to be the most important. The "PlacePics" section of Figs. 3.4 and 3.5 summarize adult and youth photovoice data.

Youth completed a total of 43 activity space maps, and adults completed a total of 21. The "PlaceTime" and "PlaceGrade" sections in Figs. 3.4 and 3.5 summarize adult and youth ASM data below (see Chap. 4 for details). Figures 3.6 and 3.7 summarize the spatial distribution of participants' combined photovoice and ASM data. Youth completed a total of 45 X-Ray Maps, while adults completed 23. The "Geography of Embodiment" section in Figs. 3.4 and 3.5 summarize adult and youth place-embodiment data across the 5 place-domains based on their completed X-Ray Maps (see Chap. 5 for details).

Data from all methods were compiled to examine what adult and youth placescapes entailed spatially, temporally, and physiologically. Figures 3.4 and 3.5 are overall graphic summaries of adult and youth placescapes based on the data they generated for all research methods. Infographics were chosen as a method to display

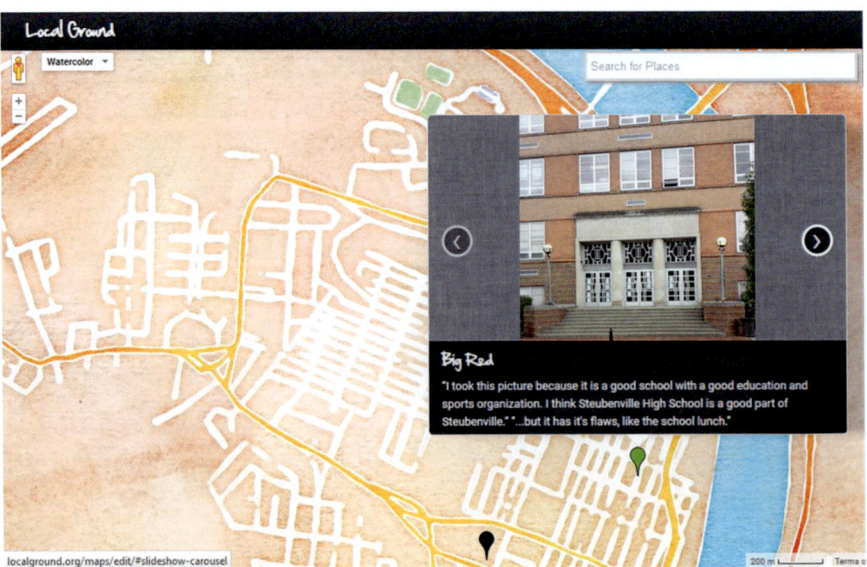

Fig. 3.2 Youth geolocated photovoice place on *Local Ground*. Photo from thematic photo group youth ranked as most important, "Positive Buildings." Participants identified 4 unique codes with a total 22 codings. The green marker corresponding to the photo indicates their appraisal of this particular place as positive/good/healthy. The black marker represents their housing community

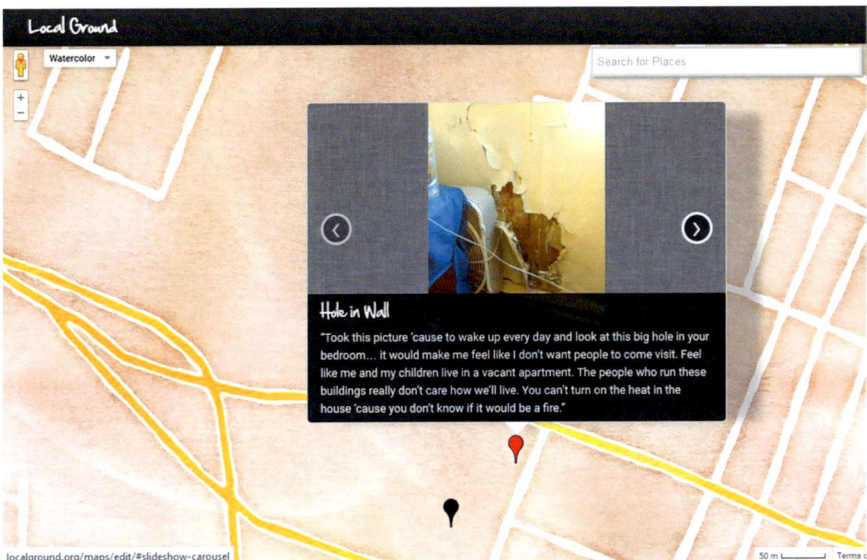

Fig. 3.3 Adult geolocated photovoice place on *Local Ground*. Photo from thematic photo group adults ranked as most important, "Housing." Participants identified 4 unique codes with a total of 10 codings. The red marker corresponding to the photo indicates their appraisal of this particular place as negative/bad/unhealthy. The black marker represents their housing community

participants' data based on discussions (led by youth participants) regarding ease of creation, use, interpretation, and integration within popular social media platforms (e.g., tagging them on Facebook, Twitter, Instagram). It should be noted that the infographics presented here are the first iterations informed and approved by the participants; however, these versions were not created by the participants' themselves as not all had sufficient time and internet access to receive basic training during the data analysis phase of this project. These infographics are based on the aggregated photovoice, ASM, and X-Ray Mapping data for adults and youth, separately. *PlacePics* represent the number of photos from photovoice across the place-domains, including only those for which place-domain categorization was possible (60 for youth, 49 for adults). *PlaceTime* represents the average estimated time participants spent across the place-domains based on their ASM data. *Geography of Embodiment* summarizes their place-embodiment perceptions across place-domains based on their X-Ray Mapping data. *PlaceGrade* is the average rating participants assigned to their various places across the place-domains based on their ASM data.

Figure 3.8 illustrates the multinodal structure of "place" for youth participants. The majority of youth data for photovoice and ASM tended to "cluster" in six particular areas of the city. Each of these areas contained at least four unique photovoice or ASM data elements (e.g., photos and activity space maps). Following the place-domain color scheme from Figs. 3.4 and 3.5 above, the green node represents participants' "Home," here accounted for by participants' in-home and immediate

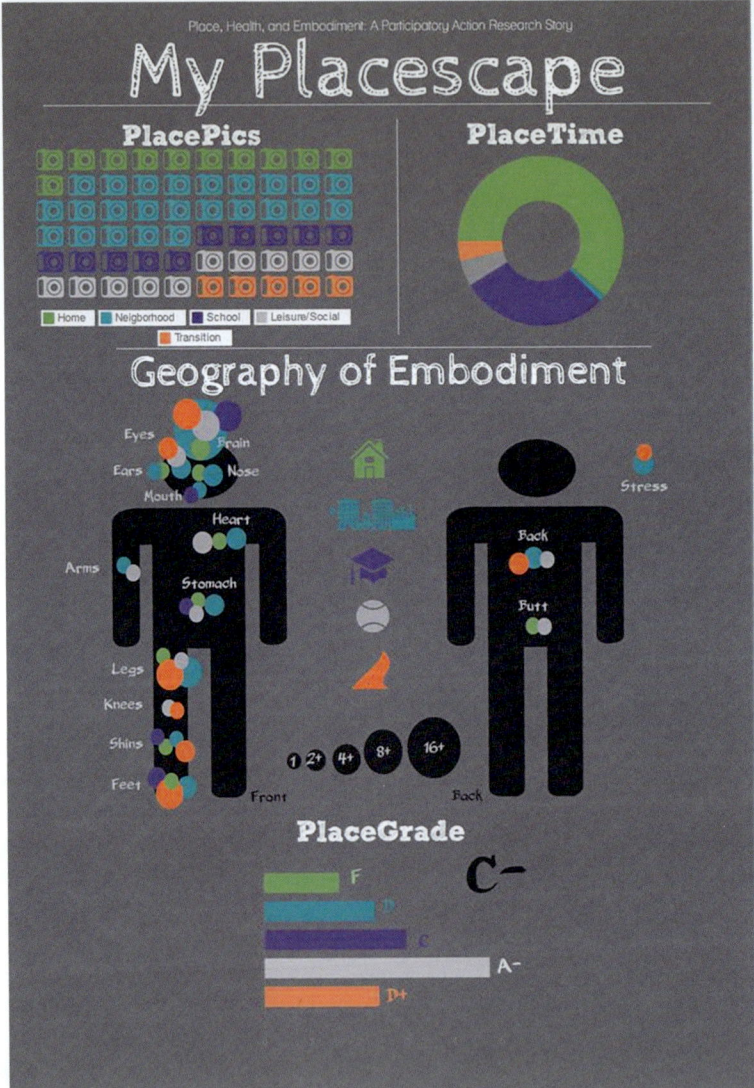

Fig. 3.4 Youth placescape summary infographic. _PlacePics_: number of participant photovoice photos for each place-domain; PlaceTime: participants' averaged estimated time spent (minutes per day) within each place-domain; _Geography of Embodiment_: participants' subjective representation of where each place-domain is physiologically embodied; _PlaceGrade_: average grade participants assigned each to place-domain based on a 5.0 scale (1 = F; 5 = A)

housing project community environment (14 data elements). The two blue nodes represent "Neighborhood," here accounted for by a cluster of vacant properties frequented when visiting friends (11 data elements), and their nearest community shopping plaza (4 data elements). The two purple nodes represent "School," here accounted for by participants' middle school (6 data elements) and high school (5

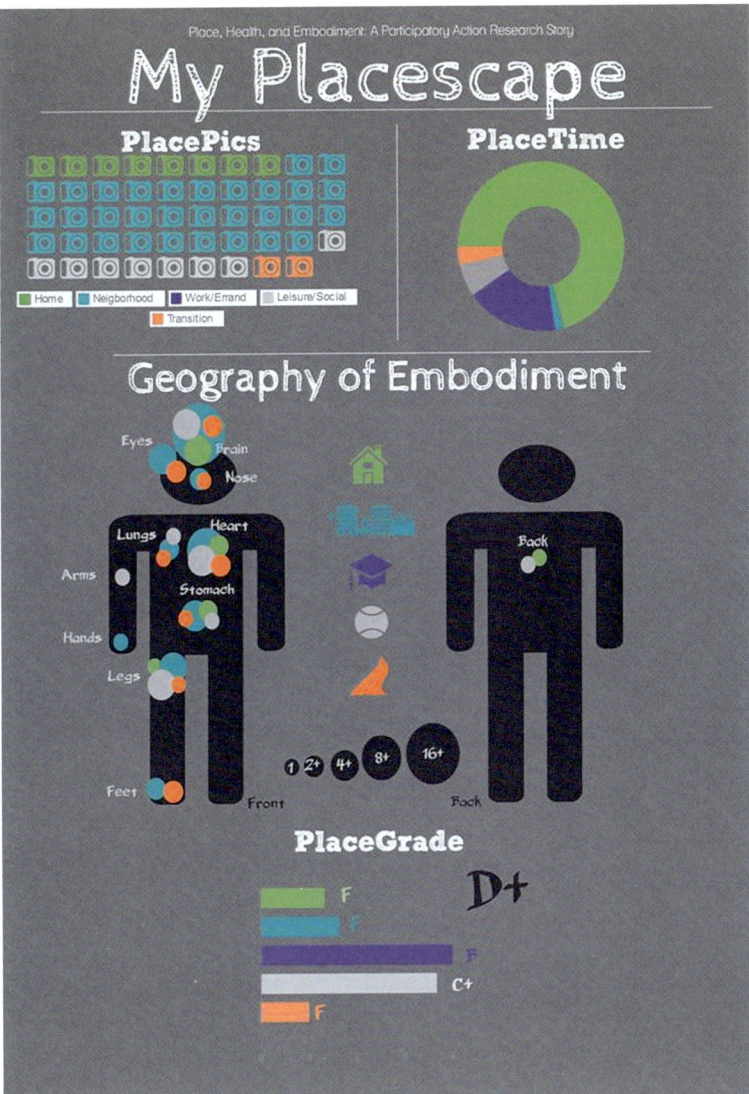

Fig. 3.5 Adult placescape summary infographic. _PlacePics_: number of participant photovoice photos for each place-domain; PlaceTime: participants' averaged estimated time spent (minutes per day) within each place-domain; Geography _of Embodiment_: participants' subjective representation of where each place-domain is physiologically embodied; _PlaceGrade_: average grade participants assigned each to place-domain based on a 5.0 scale (1 = F; 5 = A)

data elements). The grey node represents "Leisure/Social," here accounted for by an afterschool youth development center (4 data elements). Taken together, these 6 "nodes" account for roughly two-thirds (44/67) of youth participants' data elements. Notice that four of their six primary place nodes are outside their residential census tract (the black polygon outline). Figure 3.9 illustrates adults' multinodal

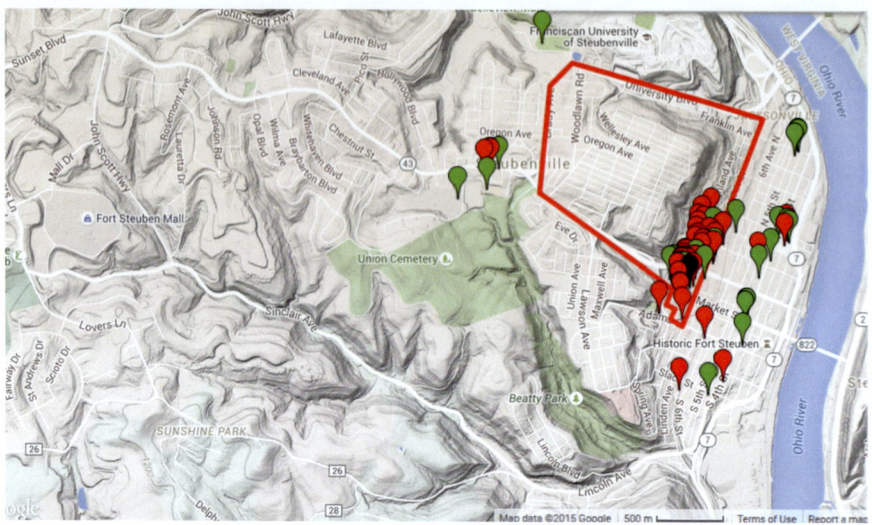

Fig. 3.6 Spatial distribution of youth photovoice and activity space mapping places. Green = positive/healthy/good place. Red = negative/unhealthy/bad place. Black marker = participants' housing community. Polygon outline = participants' residential census tract

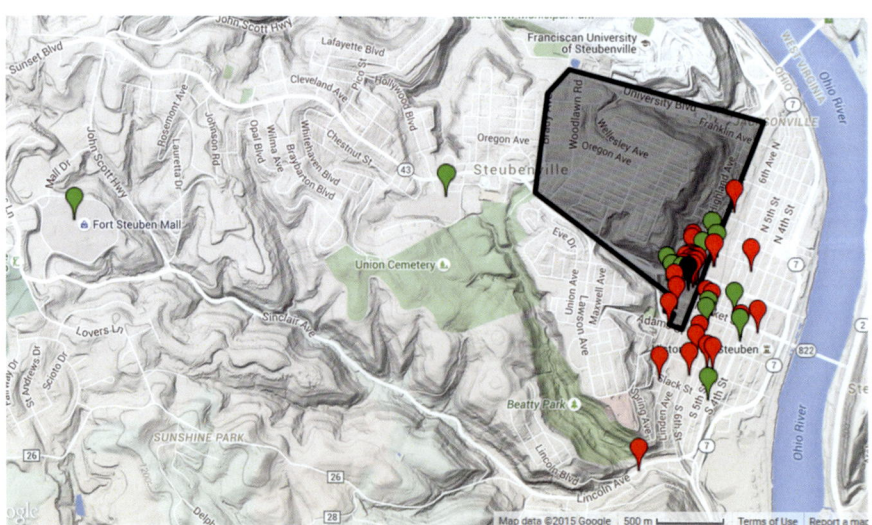

Fig. 3.7 Spatial distribution of adult photovoice and activity space mapping places. Green = positive/healthy/good place. Red = negative/unhealthy/bad place. Black marker = participants' housing community. Polygon outline = participants' residential census tract

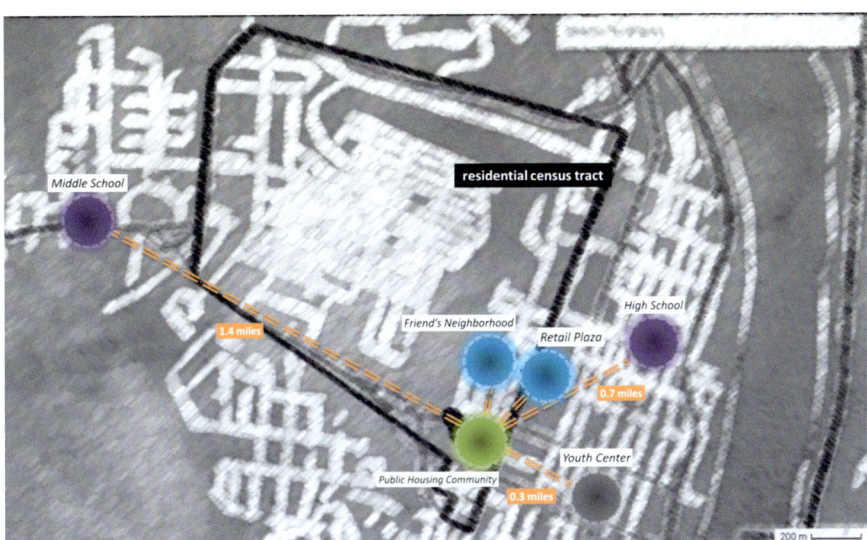

Fig. 3.8 Youth participants' "Multinodal" placescape. Illustration of youth participants' multinodal place network. Purple "nodes" represent part of their "School" place-domain; blue is "Neighborhood"; gray is "Leisure/Social"; green is "Home"

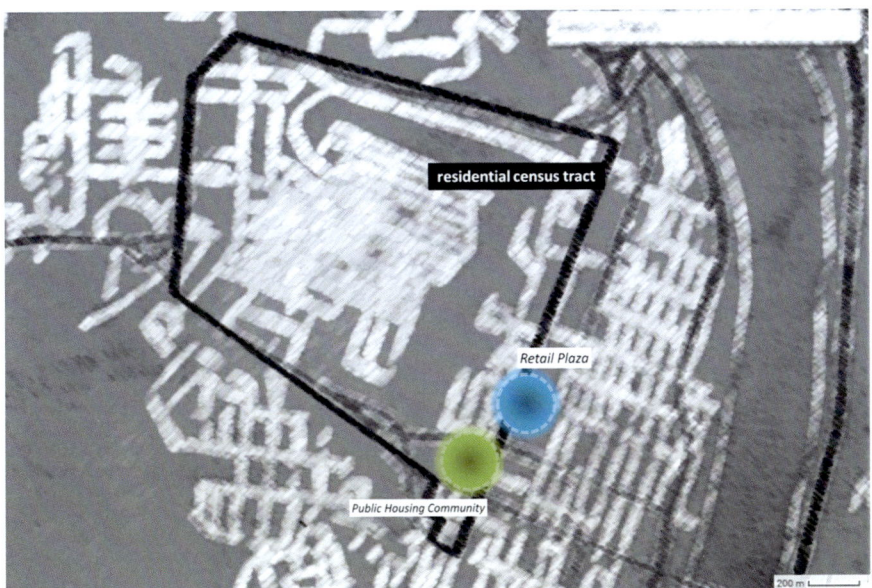

Fig. 3.9 Adult participants' "Multinodal" placescape. Illustration of adult participants' multinodal place network. The blue "node" represents part of their "Neighborhood" place-domain; green is "Home"

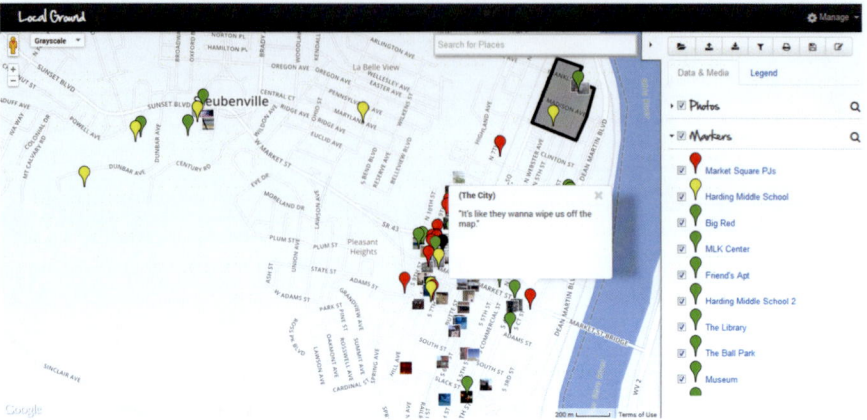

Fig. 3.10 Summary map of adult's and youth's aggregate "placescape". Green = positive/healthy/
good place. Red = negative/unhealthy/bad place. Black marker = participants' housing community.
The shaded rectangular polygon represents a walking route adult participants reported completing
frequently, about 1 mile from their residential community

"place," with markedly fewer nodes in comparison to youth. Adults' photovoice and
ASM results revealed two clusters of data elements. The first represents part of their
"Home" place-domain (11 data elements). The second represents a part of their
"Neighborhood" place-domain (4 data elements), here consisting of the same clus-
ter of retail shops youth identified. Figure 3.10 shows a summary of aggregated
youth and adult "placescape" data across methods.

Resistance

Discussion: Participatory Placescapes as Resistance

It's like they wanna wipe us off the map. *Adult Participant*

The goal of the work presented here was to introduce the *placescape* as an analytical framework for understanding place and health in the context of place-based
strategies and to outline a conceptual foundation for operationalizing it in the context of place-health research. Taking a placescape approach to studying place and
health can enhance our conceptual understanding of place, and can consequently
inform and improve place-health research design, metrics, and methodological
choices, as well as the development and evaluation of place-based strategies.
Importantly, it can offer a conceptual grounding toward countering and decolonizing representations of public health/housing geographies and dominant spatial narratives of health thereof. The work pursued here attempted to operationalize the six
core *placescape* tenets (PTs) from Table 3.2 in a "field test" to aid efforts on each of
these fronts.

In regard to *Needs & Opportunities* (PT1), *Mobility & Bounds* (PT2), *Multinodal Place* (PT3), and *Agency in Place(making)* (PT6), the participatory and non-arbitrary, non-spatially bound approach of this project allowed participants to freely document their important daily places as they actually experience them. This not only facilitated revelation of what could be considered some of participants' basic daily necessities (e.g., education, retail access, social/recreation spaces) but also where these necessities were spatially located and how participants perceived they influenced health (Figs. 3.6 and 3.7). For example, a key finding here is that 55% of adult and 51% of youth places were located spatially outside of their residential census tract (data shown in Chap. 4). Indeed for youth, 80% of their positive/healthy/good places were outside of their census tract, while 67% of their negative/unhealthy/bad places were inside. Here, participant mobility, and the manner in which they repeatedly cross various imaginary lines of "place," was acknowledged and duly accounted for. Accordingly, participants' "spatial polygamy" was duly appreciated (Matthews, 2011; Matthews & Yang, 2013), which enabled the discernment of spatially diverse patterns of place-experience clustering, or place "nodes"—thus hinting at the multi-nodal structure of participants' lived place (Fig. 3.8). Moreover, because this project entailed multiple participatory methods designed to elicit participants' subjective appraisals of their place-based experiences and exposures, each place within these "nodes" (and the nodes themselves) helps "map" geographically and physiologically (via X-Ray Mapping) the health risk *and* opportunity landscape (Figs. 3.6 and 3.7), thus embracing and extending the notion of "riskscape" (Morello-Frosch et al., 2001; Morello-Frosch & Shenassa, 2006). For example, participant X-Ray Mapping data revealed that 62% of adult and 49% of youth place-embodiment reports were for places spatially outside of their residential census tract (see Chap. 5). An even more important finding was that 75% of youth positive place-embodiment places were located outside their census tract, while 66% of the negative place-embodiment locations were located inside. Indeed overall, Figs. 3.6 and 3.7 make it clear that, spatially speaking, the overwhelming majority of participants' census tract has no bearing on their place experiences. These findings lend further support to literature raising concerns over the mis-specification of place-effects (Cummins, 2007; Inagami et al., 2007; Kwan, 2009, 2012; Mujahid et al., 2007; Riva et al., 2008; Roux, 2007; Spielman & Yoo, 2009) and calling for more nuanced approaches to studying place and health (Browning & Soller, 2014; Cummins et al., 2007; Cutchin et al., 2011; Jones & Pebley, 2014; Macintyre et al., 2002; Rainham et al., 2010) (Macintyre et al., 2002; Cummins et al., 2007; Rainham et al., 2010; Cutchin et al. 2011; Browning and Soller, 2014; Jones and Pebley, 2014). The placescape approach as outlined and applied here stands as a conceptually rich prism through which to examine work that is beginning to engage place beyond residential areas but is still based on administrative boundaries (e.g., Hoehner et al., 2013; Moore et al., 2013).

In regard to *Lifecourse in Place* (PT5), the intergenerational design of this project meant that both adult and youth perspectives were engaged. This allowed for an exploration of not only potential spatial differences between adult and youth places, but also their perceptual differences of place and its positive and negative embodied health effects (Figs. 3.4, 3.5, 3.6 and 3.7). Accordingly, this project responds to calls for lifecourse perspectives in place-health research (Curtis et al., 2004; Gustafsson et al., 2014; Pearce, 2015) and makes a rare qualitative contribution to a growing body of literature regarding the physiological embodiment of place over time (Bird et al., 2010; Brenner et al., 2013; Broyles et al., 2012; Crimmins et al., 2003; Gustafsson et al., 2011; King et al., 2011; Merkin et al., 2009; Nazmi et al., 2010; Petersen et al., 2008; Petteway et al., 2019a; Rudolph et al., 2014; Theall et al., 2012). Relatedly, and in regard to *Mobility & Bounds* (PT2), *Multinodal Place* (PT3), and *Lifecourse in Place* (PT5), this project attempted to explore and account for temporality in the context of place experiences/exposures. Temporality, here, was focused on elucidating participants' place-specific time patterns. The goal was to get a greater sense of which places (or nodes) tend to account for the most place "exposure time," with the goal of being able to outline ways to time-weight place experiences. The Activity Space Mapping process employed for this project proved useful here (Figs. 3.4 and 3.5), if only in an exploratory and introductory sense. Even so, this project does well to illustrate potential ways to arrive at both spatially and temporally specific measures of place—and how to do so in a manner that is participatory and sensitive to potential life-stage or generation-contingent differences. Thus, this work moves us further along in our pursuit of more dynamic and specific measures of place (Chaix et al., 2009; Cummins et al., 2005; Mujahid et al., 2007; O'Campo, 2003; Roux, 2007).

In relation to *Lifecourse in Place* (PT5) and *Agency in Place(making)* (PT6), this project employed only participatory action research methods with two generations of public housing residents. The results accordingly tell a story of place and health that traditional approaches tend to overlook and discount. By adopting a CBPR orientation, and an intergenerational one at that, this project was able to uncover place-health nuances not readily afforded to standard survey-based, administrative-boundary delimited research. Of important note here is that this work was greatly facilitated through the use of a collaborative web-based and multimedia-enabled information and communication technology (ICT) platform, *Local Ground*. The result was location-specific, time-specific, and generation-specific illustrations of place and place perceptions that were just as visceral as they are empirical, and in their production served as a form of social and political expression—their places-capes could be readily shared with one another, via social media, and with city officials (touching on *Power in Place(making)*). This more thorough and embodied rendition of place holds specific value not only in the context of public health practice and place-based strategizing (e.g., as a model for community assessment and participatory health and city planning), but also in the context of place-health research translation. Moreover, the work presented here encourages a more focused exploration into the value and relative importance of "objective" and "subjective" measures of place. Place-health research to date has demonstrated that both matter (Barrington et al., 2014; Lin & Moudon, 2010; Pampalon et al., 2007; Schulz et al., 2013; van Deurzen et al., 2016; Weden et al., 2008; Wen et al., 2006), and indeed, work involving public housing residents suggests that subjective place data may be more meaningful and predictive of well-being (Buron & Patrabansh, 2008). The participatory approach taken in this work illustrates a decolonized orientation to studying place-health relationships and the creation of representations/narratives thereof. In doing so, it centers on the importance of agency and power in knowledge production about place, thereby acknowledging that *researching and communicating about place* and health is a fundamental element of how place continues to be (re)made. In this context, spatial dispossession must be understood not only in historic and literal terms, but in present, future, and figurative terms, as knowledge representations and data narratives of place produced via extractivist and colonized processes serve the same function—to subjugate, own, exploit, and capitalize.

The work presented here also highlights possibilities for more innovative mixed-methods examinations of place and health, e.g., via utilization of ICTs and GIS (Chaix et al., 2012; Dennis et al., 2009; Fielding & Cisneros-Puebla, 2009; Kwan & Ding, 2008). *Local Ground* served as a mechanism to "crowdsource" local community assessment and enhance resident voice via potential linkages with the local health department and planning commission. This work could potentially serve as a model specifically for participatory planning and community health/opportunity assessment within public housing and other HUD place-based strategies (Petteway, 2021). For example, this sort of work could be used to systematically document and assess specific community housing conditions, both social and physical, the results of which could be used to inform larger scale (i.e., entire housing community) survey-based quantitative or mixed-methods (e.g., surveys, photo/video, GIS) efforts. At minimum, HUD public housing communities and local housing authorities could use this work as a model to develop and execute regular "community health" assessments among and *with* residents, the findings of which can be examined in relation to, and be used to supplement, the standard HUD housing inspection process (HUD, 2015g, h), as well as the Healthy Communities Assessment Tool (HUD, 2015b).

Lastly, in light of *Power in Place(making)* (PT4) and *Agency in Place(making)* (PT6), this project engaged residents of public housing through a CBPR approach, using exclusively participatory methods. As such, residents were able to identify and reveal their place-related perspectives, concerns, and experiences from a position of expertise in a process marked by mutual respect and co-learning. They did so not only through their mapping-oriented work but also during group training and project meeting sessions. For example, in the context of a discussion regarding health in public housing and in light of what residents believe is the city's deliberate effort to get rid of public housing, one adult participant submitted that "it's like they wanna wipe us off the map"—both poignant and apropos given the nature of the project. "They" referred to city officials whom the participants believe are "configuring" their places to an overwhelming deleterious effect—ostensibly allowing conditions to deteriorate to the extent that their housing communities are condemned, or extreme un-inhabitability forces a move—a practice discussed in detail in Goetz (2013a). Indeed, that is precisely what had happened to at least three low-income housing communities in recent years. One such community was adjacent to a prestigious, predominantly White private Catholic university. Relations between mostly Black residents and the university were tenuous to say the least, and city officials agreed to sell the property to the university after allowing it to deteriorate—a rather blatant and direct example of displacement and accumulation by dispossession that was unmistakably racialized.

Moving for the project participants, however, seems increasingly improbable, perhaps impossible given the increasing deficit of affordable housing options. While the city recently released its new comprehensive plan, there is no clear indication of plans regarding public housing in participants' downtown community, and none of them were invited to inform the development of the plan (or were aware of it, for that matter). Making matters worse, a recently completed housing development—the only new multiunit complex developed in over 20 years—had been reserved exclusively for students of the university. Not only that, but there was a growing sense of housing discrimination among the adult participants. As one participant put it in describing her attempts to find new housing, "they find out where you live and they don't even want you to apply." The stigma attached to her as a tenant of her current housing project—stigma in part due to the city's failure to adequately maintain the property and support its tenants—prevented her from being considered an acceptable tenant elsewhere, reflecting an element of what Smith (2005) refers to as "selective entrapment." The many policies and practices (actions and inactions) that have historically shaped and continue to influence residents' configurations of place nodes, of course, need to be more thoroughly expounded. Nonetheless, the aggregate work completed by the participants here stands as a good starting point to unpack these larger issues, particularly in regard to notions of dispossession, agency, accountability, and processes of inclusion and exclusion within local (public) housing practice.

In a more immediate sense, the results here highlight the heterogeneity of "place" perceptions and experiences among people who share the same space, with a few key take-away implications for place-based strategies in the city. First, participants' work revealed a clear pattern of positive/negative places to suggest where to invest more, and possibly where to best locate new housing. Second, there was a clear indication of wasted/vacant space frequently encountered and traversed by residents—participants' work suggested where community enhancements could be made (greenspace, urban gardens, sports fields, new retail and cafes, public transportation improvements, etc.), where there might be physical space for new housing, and where there might be clustered community safety concerns. Third, there was a clear indication of where residents spend their time and social lives—hinting at the structure and spatial bounds of their social lives, and suggesting that their sense of belonging and social embeddedness, both space-wise and community-wise, is "downtown." These findings suggest that locating potential new affordable housing elsewhere in the city is not appropriate (again, see Figs. 3.6 and 3.7), and that doing so would ignore the community's rootedness and potentially function to disrupt their social ties. While this potentially "luring" approach to community destabilization is not as direct as displacement mechanisms like demolition and gentrification, it may still cause/contribute to what Fullilove (2001, 2004) refers to as "root shock"—generally, traumatic stress and sociopolitical fragmentation resultant from disrupting social bonds and emotional ties to place.

In aggregate, the work presented here contributes to the field of place-health research, as well as place-based public health and community development practice and housing strategies. Even so, there are a few limitations and key areas to improve upon in future work. First, this work did not adequately capture notions related to accountability and agency in terms of participants' perceptions on who is responsible and who has power in influencing where they live/don't live, where they go/ don't go, etc. More qualitative information on these aspects of their placescapes would be valuable. Second, and relatedly, participant training and methods for this project did not allow for the explicit identification and enumeration of fundamental material placemaking mechanisms at the local level, i.e., actual local/regional housing and land use policies, or the role of the CDBG and LIHTC locally. For example, it would have been valuable to track the housing and community development history in relation to how low-income housing funds had been allocated and how siting decisions were made, e.g., how it came to be that public housing communities were situated next to major roads, train tracks, and industrial facilities. Collaboratively gathering more information on actual local public housing and community development policies and practices, both currently and historically, could prove especially valuable in uncovering the determinants and structure of residents' lived placescapes—allowing for more explicit accounting of what Smith (2005) calls "selective placement," and the manner in which racialization of space governs where residents can live and how investments are determined therefrom. This could be a particularly promising endeavor to be pursued through follow-up CBPR and youth participatory action research work.

Third, this project did not adequately capture spatial restrictions/divisions or affinities within residents' configurations of place nodes, i.e., examining the time-space geography notion of *authority* (Hägerstrand, 1970), and notions of *belonging* or *place attachment* (Leung & Takeuchi, 2011). As such, there is a need for more qualitative and geospatial information on the places "missing" from their placescapes—understanding why they do not go to certain places is just as much a part of their placescape as the places they do go. These unspecified places are a sort of "invisible placescape" that structures their lived placescapes—for example, when not feeling particularly safe in a neighborhood influences daily walking routes, when not feeling welcomed at certain parks or stores, when feeling "surveilled" in certain neighborhoods or community spaces. Future work could better characterize the role of structural factors like racial and economic segregation, policing practices, infrastructure disinvestment, and city planning/design decisions in order to acknowledge/appreciate places they did not identify because they *can't* go there. Relatedly, future work could better elucidate and characterize spaces of inclusion, support, and belonging. Fourth, the methods as implemented did not adequately capture the timing and temporal ordering of participants' places and place-experiences, e.g., in the morning, only in the evening, before this, after that, during this, only in the summer. Part of this limitation is simply the nature of cross-sectional work and a time-intensive participatory approach, but follow-up efforts would be enhanced greatly by exploring more systematic ways to infuse participants' placescapes with a greater sense of time and timing. Applying more thorough and robust quantitative activity space and/or qualitative GIS approaches could prove particularly helpful here (Chaix et al., 2012; Jung & Elwood, 2010; Kwan & Ding, 2008) (e.g., Chaix et al., 2012; Kwan and Ding, 2008; Jung and Elwood, 2010), perhaps making use of ecological momentary assessment and/or citizen science approaches (Collins et al., 2003; Conrad & Hilchey, 2011; Den Broeder et al., 2016; Dunton et al., 2012; English et al., 2018; Shiffman et al., 2008; Spook et al., 2013).

Conclusion

A participatory *placescape* approach to studying place and health can enhance our conceptual understanding of place and consequently inform and improve place-health research metrics and methodological potentials. Moreover, a placescape orientation could enhance prospects for research translation in ways that can make direct contributions to local/regional public health and city planning/community development practice. As applied here, this approach can characterize the diversity of place-based experiences among public housing residents, yielding spatially and generationally specific information that can inform evaluation of current housing conditions and strategies, as well as guide the design and implementation of future strategies, especially in relation to matters of health equity and place-based health opportunity. More fundamentally, this approach can facilitate the decolonization of place-health narratives and the sociospatial (mis)representations of residents' daily lives, affording greater opportunities to locate placemaking as (re)productive of social and material contexts of place-health relationships—in public housing or otherwise.

References

Acevedo-Garcia, D. (2004). Does housing mobility policy improve health? *Housing Policy Debate, 15*(1), 49–98.

Arcaya, M. C., Tucker-Seeley, R. D., Kim, R., Schnake-Mahl, A., So, M., & Subramanian, S. V. (2016). Research on neighborhood effects on health in the United States: A systematic review of study characteristics. *Social Science & Medicine, 168*, 16–29. https://doi.org/10.1016/j.socscimed.2016.08.047

Barrington, W. E., Stafford, M., Hamer, M., Beresford, S. A. A., Koepsell, T., & Steptoe, A. (2014). Neighborhood socioeconomic deprivation, perceived neighborhood factors, and cortisol responses to induced stress among healthy adults. *Health & Place, 27*, 120–126. https://doi.org/10.1016/j.healthplace.2014.02.001

Bird, C. E., Seeman, T., Escarce, J. J., Basurto-Davila, R., Finch, B. K., Dubowitz, T., Heron, M., Hale, L., Merkin, S. S., Weden, M., & Lurie, N. (2010). Neighbourhood socioeconomic status and biological "wear and tear" in a nationally representative sample of US adults. *Journal of Epidemiology & Community Health, 64*(10), 860–865. https://doi.org/10.1136/jech.2008.084814

Bloom, N. D., Umbach, F., & Vale, L. J. (Eds.). (2015). *Public housing myths: Perception, reality, and social policy* (Illustrated edn.). Cornell University Press.

Brenner, A. B., Zimmerman, M. A., Bauermeister, J. A., & Caldwell, C. H. (2013). Neighborhood context and perceptions of stress over time: An ecological model of neighborhood stressors and intrapersonal and interpersonal resources. *American Journal of Community Psychology, 51*(3–4), 544–556. https://doi.org/10.1007/s10464-013-9571-9

Browning, C. R., & Soller, B. (2014). Moving beyond neighborhood: Activity spaces and ecological networks as contexts for youth development. *Cityscape (Washington, DC), 16*(1), 165.

Broyles, S. T., Staiano, A. E., Drazba, K. T., Gupta, A. K., Sothern, M., & Katzmarzyk, P. T. (2012). Elevated C-reactive protein in children from risky neighborhoods: Evidence for a stress pathway linking neighborhoods and inflammation in children. *PLoS ONE, 7*(9), e45419. https://doi.org/10.1371/journal.pone.0045419

Buron, L., & Patrabansh, S. (2008). Are census variables highly correlated with Housing Choice Voucher holders' perception of the quality of their neighborhoods? *Cityscape, 10*, 157–183.

Buron, L., Comey, J., Cunningham, M. K., Harris, L. E., Levy, D., & Popkin, S. J. (2002). *HOPE VI panel study: Baseline report* (p. 283). The Urban Institute.

Cairns, K. (2018). Youth, temporality, and territorial stigma: Finding good in Camden, New Jersey. *Antipode, 50*(5), 1224–1243. https://doi.org/10.1111/anti.12407

Catalani, C., & Minkler, M. (2010). Photovoice: A review of the literature in health and public health. *Health Education & Behavior, 37*(3), 424–451. https://doi.org/10.1177/1090198109342084

Chaix, B., Merlo, J., Evans, D., Leal, C., & Havard, S. (2009). Neighbourhoods in eco-epidemiologic research: Delimiting personal exposure areas. A response to Riva, Gauvin, Apparicio and Brodeur. *Social Science & Medicine, 69*(9), 1306–1310. https://doi.org/10.1016/j.socscimed.2009.07.018

Chaix, B., Kestens, Y., Perchoux, C., Karusisi, N., Merlo, J., & Labadi, K. (2012). An interactive mapping tool to assess individual mobility patterns in neighborhood studies. *American Journal of Preventive Medicine, 43*(4), 440–450. https://doi.org/10.1016/j.amepre.2012.06.026

Chaskin, R. J. (2013). Integration and exclusion: Urban poverty, public housing reform, and the dynamics of neighborhood restructuring. *The Annals of the American Academy of Political and Social Science, 647*(1), 237–267. https://doi.org/10.1177/0002716213478548

Clampet-Lundquist, S. (2004a). Moving over or moving up? Short-term gains and losses for relocated HOPE VI families. *Cityscape, 7*(1), 57–80.

Clampet-Lundquist, S. (2004b). HOPE VI relocation: Moving to new neighborhoods and building new ties. *Housing Policy Debate, 15*(2), 415–447. https://doi.org/10.1080/10511482.2004.9521507

Collins, R. L., Kashdan, T. B., & Gollnisch, G. (2003). The feasibility of using cellular phones to collect ecological momentary assessment data: Application to alcohol consumption. *Experimental and Clinical Psychopharmacology, 11*(1), 73–78. https://doi.org/10.1037/1064-1297.11.1.73

Conrad, C. C., & Hilchey, K. G. (2011). A review of citizen science and community-based environmental monitoring: Issues and opportunities. *Environmental Monitoring and Assessment, 176*(1–4), 273–291. https://doi.org/10.1007/s10661-010-1582-5

Crimmins, E. M., Johnston, M., Hayward, M., & Seeman, T. (2003). Age differences in allostatic load: An index of physiological dysregulation. *Experimental Gerontology, 38*(7), 731–734. https://doi.org/10.1016/S0531-5565(03)00099-8

Cummins, S. (2007). Commentary: Investigating neighbourhood effects on health–Avoiding the "Local Trap". *International Journal of Epidemiology, 36*(2), 355–357. https://doi.org/10.1093/ije/dym033

Cummins, S., Macintyre, S., Davidson, S., & Ellaway, A. (2005). Measuring neighbourhood social and material context: Generation and interpretation of ecological data from routine and non-routine sources. *Health & Place, 11*(3), 249–260. https://doi.org/10.1016/j.healthplace.2004.05.003

Cummins, S., Curtis, S., Diez-Roux, A. V., & Macintyre, S. (2007). Understanding and representing 'place' in health research: A relational approach. *Social Science & Medicine, 65*(9), 1825–1838. https://doi.org/10.1016/j.socscimed.2007.05.036

Curtis, S., Southall, H., Congdon, P., & Dodgeon, B. (2004). Area effects on health variation over the life-course: Analysis of the longitudinal study sample in England using new data on area of residence in childhood. *Social Science & Medicine, 58*(1), 57–74. https://doi.org/10.1016/S0277-9536(03)00149-7

Cutchin, M. P., Eschbach, K., Mair, C. A., Ju, H., & Goodwin, J. S. (2011). The socio-spatial neighborhood estimation method: An approach to operationalizing the neighborhood concept. *Health & Place, 17*(5), 1113–1121. https://doi.org/10.1016/j.healthplace.2011.05.011

de Souza Briggs, X. (2005). *The geography of opportunity: Race and housing choice in Metropolitan America.* Brookings Institution Press. https://www.jstor.org/stable/10.7864/j.ctt1gpccgb

Delgado, R., Stefancic, J., & Harris, A. (2017). *Critical race theory* (3rd ed.). NYU Press.

Den Broeder, L., Devilee, J., Van Oers, H., Schuit, A. J., & Wagemakers, A. (2016). Citizen Science for public health. *Health Promotion International, 33*, daw086. https://doi.org/10.1093/heapro/daw086

Dennis, S. F., Gaulocher, S., Carpiano, R. M., & Brown, D. (2009). Participatory photo mapping (PPM): Exploring an integrated method for health and place research with young people. *Health & Place, 15*(2), 466–473. https://doi.org/10.1016/j.healthplace.2008.08.004

Diez Roux, A. V., & Mair, C. (2010). Neighborhoods and health: Neighborhoods and health. *Annals of the New York Academy of Sciences, 1186*(1), 125–145. https://doi.org/10.1111/j.1749-6632.2009.05333.x

Digenis-Bury, E. C., Brooks, D. R., Chen, L., Ostrem, M., & Horsburgh, C. R. (2008). Use of a population-based survey to describe the health of boston public housing residents. *American Journal of Public Health, 98*(1), 85–91. https://doi.org/10.2105/AJPH.2006.094912

Dunton, G. F., Kawabata, K., Intille, S., Wolch, J., & Pentz, M. A. (2012). Assessing the social and physical contexts of children's leisure-time physical activity: An ecological momentary assessment study. *American Journal of Health Promotion, 26*(3), 135–142. https://doi.org/10.4278/ajhp.100211-QUAN-43

Edwards, F. L., & Thomson, G. B. (2010). The legal creation of raced space: The subtle and ongoing discrimination created through Jim Crow laws. *Berkeley Journal of African-American Law and Policy, 12*, 145.

Ellen, I. G., Mijanovich, T., & Dillman, K.-N. (2001). Neighborhood effects on health: Exploring the links and assessing the evidence. *Journal of Urban Affairs, 23*(3–4), 391–408.

English, P. B., Richardson, M. J., & Garzón-Galvis, C. (2018). From crowdsourcing to extreme citizen science: Participatory research for environmental health. *Annual Review of Public Health, 39*(1), 335–350. https://doi.org/10.1146/annurev-publhealth-040617-013702

Fauth, R. (2008). Seven years later: Effects of a neighborhood mobility program on poor black and Latino adults' well-being. *Journal of Health and Social Behavior, 49*(2), 119–130.

FDIC. (2015). *Federal deposit insurance corporation: Community reinvestment act program site of the Federal Deposit Insurance Corporation.* https://www.fdic.gov/regulations/cra/

Fertig, A. R., & Reingold, D. A. (2007). Public housing, health, and health behaviors: Is there a connection? *Journal of Policy Analysis and Management, 26*(4), 831–860. https://doi.org/10.1002/pam.20288

Fielding, N., & Cisneros-Puebla, C. A. (2009). CAQDAS-GIS convergence: Toward a new integrated mixed method research practice? *Journal of Mixed Methods Research, 3*(4), 349–370. https://doi.org/10.1177/1558689809344973

Fine, M., & Ruglis, J. (2009). Circuits and consequences of dispossession: The racialized realignment of the public sphere for U.S. youth. *Transforming Anthropology, 17*(1), 20–33. https://doi.org/10.1111/j.1548-7466.2009.01037.x

Ford, C. L., & Airhihenbuwa, C. O. (2010). The public health critical race methodology: Praxis for antiracism research. *Social Science & Medicine, 71*(8), 1390–1398. https://doi.org/10.1016/j.socscimed.2010.07.030

Foucault, M., & Senellart, M. (2008). *The birth of biopolitics: Lectures at the Collège de France, 1978-79.* Palgrave Macmillan.

Fullilove, M. T. (2001). Root shock: The consequences of African American dispossession. *Journal of Urban Health, 78*(1), 72–80.

Fullilove, M. (2004). *Root shock: How tearing up city neighborhoods Hurts America, and what we can do about it.* New Village Press. https://nyupress.org/9781613320198/root-shock

Galster, G. C., & Killen, S. P. (1995). The geography of metropolitan opportunity: A reconnaissance and conceptual framework. *Housing Policy Debate, 6*(1), 7–43. https://doi.org/10.1080/10511482.1995.9521180

Galster, G., & Sharkey, P. (2017). Spatial foundations of inequality: A conceptual model and empirical overview. *RSF: The Russell Sage Foundation Journal of the Social Sciences, 3*(2), 1. https://doi.org/10.7758/rsf.2017.3.2.01

Goetz, E. G. (2010). Better neighborhoods, better outcomes? Explaining relocation outcomes in Hope VI. *SSRN Electronic Journal.* https://doi.org/10.2139/ssrn.1585369

Goetz, E. (2013a). *New deal ruins: Race, economic justice, and public housing policy.* . https://www.cornellpress.cornell.edu/book/9780801478284/new-deal-ruins/

Goetz, E. (2013b). The audacity of HOPE VI: Discourse and the dismantling of public housing. *Cities, 35*, 342–348.

Goetz, E. G., & Chapple, K. (2010). You gotta move: Advancing the debate on the record of dispersal. *Housing Policy Debate, 20*(2), 209–236. https://doi.org/10.1080/10511481003779876

Golledge, R., & Stimson, R. (1996). *Spatial behavior: A geographic perspective.* Guilford Press. https://www.guilford.com/books/Spatial-Behavior/Golledge-Stimson/9781572300507

Graham, L. F., Padilla, M. B., Lopez, W. D., Stern, A. M., Peterson, J., & Keene, D. E. (2016). Spatial stigma and health in postindustrial detroit. *International Quarterly of Community Health Education, 36*(2), 105–113. https://doi.org/10.1177/0272684X15627800

Gustafsson, P. E., Janlert, U., Theorell, T., Westerlund, H., & Hammarstrom, A. (2011). Socioeconomic status over the life course and allostatic load in adulthood: Results from the Northern Swedish Cohort. *Journal of Epidemiology & Community Health, 65*(11), 986–992. https://doi.org/10.1136/jech.2010.108332

Gustafsson, P. E., San Sebastian, M., Janlert, U., Theorell, T., Westerlund, H., & Hammarström, A. (2014). Life-course accumulation of neighborhood disadvantage and allostatic load: Empirical integration of three social determinants of health frameworks. *American Journal of Public Health, 104*(5), 904–910.

Hägerstrand, T. (1970). What about people in regional science? *Papers of the Regional Science Association, 24*(1), 6–21. https://doi.org/10.1007/BF01936872

Halliday, E., Popay, J., Anderson de Cuevas, R., & Wheeler, P. (2020). The elephant in the room? Why spatial stigma does not receive the public health attention it deserves. *Journal of Public Health (Oxford, England), 42*(1), 38–43. https://doi.org/10.1093/pubmed/fdy214

Harris, L. E., & Kaye, D. R. (2004). *How are HOPE VI families faring? health.* .

Harvey, D. (2004). The "New" imperialism: Accumulation by dispossession. *Socialist Register, 40.* https://socialistregister.com/index.php/srv/article/view/5811

Harvey, D. (2006). *Spaces of global capitalism: A theory of uneven geographical development* (1st ed.). Verso.

HCZ. (n.d.). *Harlem children's zone.* Retrieved February 27, 2013, from http://hcz.org/index.php

Hoehner, C. M., Allen, P., Barlow, C. E., Marx, C. M., Brownson, R. C., & Schootman, M. (2013). Understanding the independent and joint associations of the home and workplace built environments on cardiorespiratory fitness and body mass index. *American Journal of Epidemiology, 178*(7), 1094–1105. https://doi.org/10.1093/aje/kwt111

Howell, E., Harris, L. E., & Popkin, S. J. (2005). The health status of HOPE VI public housing residents. *Journal of Health Care for the Poor and Underserved, 16*(2), 273–285. https://doi.org/10.1353/hpu.2005.0036

HUD. (2013a). *US Department of Housing and Urban Development: Choice neighborhoods program.* http://portal.hud.gov/hudportal/HUD?src=/program_offices/public_indian_housing/programs/ph/cn

HUD. (2013b). *US Department of Housing and Urban Development: HOPE VI program.* http://portal.hud.gov/hudportal/HUD?src=/program_offices/public_indian_housing/programs/ph/hope6/about

HUD. (2013c). *US Department of Housing and Urban Development: Resident characteristics report*. http://portal.hud.gov/hudportal/HUD?src=/program_offices/public_indian_housing/systems/pic/50058/rcr on 1/31/2013

HUD. (2013d). *US Department of Housing and Urban Development: Sustainable communities regional planning program*. http://portal.hud.gov/hudportal/HUD?src=/program_offices/sustainable_housing_communities/sustainable_communities_regional_planning_grants

HUD. (2013e). *US Department of Housing and Urban Development: Sustainable Housing and Communities program*. http://portal.hud.gov/hudportal/HUD?src=/program_offices/sustainable_housing_communities

HUD. (2015a). *US Department of Housing and Urban Development: Community Development Block Grant Entitlement Program*. https://www.hudexchange.info/programs/cdbg-entitlement/

HUD. (2015b). *US Department of Housing and Urban Development: Healthy communities assessment tool, San Diego*. https://hci-sandiego.icfwebservices.com/

HUD. (2015c). *US Department of Housing and Urban Development: Healthy communities transformation initiative*. http://healthyhousingsolutions.com/service/applied-field-research/hud-healthy-communities-transformation-initiative/

HUD. (2015d). *US Department of Housing and Urban Development: Home Investment Partnership Program (HOME)*. https://www.hudexchange.info/home/home-overview/

HUD. (2015e). *US Department of Housing and Urban Development: Housing choice voucher (Section 8) program*. http://portal.hud.gov/hudportal/HUD?src=/program_offices/public_indian_housing/programs/hcv/about

HUD. (2015f). *US Department of Housing and Urban Development: Low-Income Housing Tax Credit (LIHTC) program*. http://www.huduser.gov/portal/datasets/lihtc.html

HUD. (2015g). *US Department of Housing and Urban Development: Physical inspection summary report*. http://portal.hud.gov/hudportal/HUD?src=/program_offices/public_indian_housing/reac/products/pass/pass_isrpt

HUD. (2015h). *US Department of Housing and Urban Development: Real Estate Assessment Center, physical inspection*. http://portal.hud.gov/hudportal/HUD?src=/program_offices/public_indian_housing/reac/products/prodpass

Inagami, S., Cohen, D. A., & Finch, B. K. (2007). Non-residential neighborhood exposures suppress neighborhood effects on self-rated health. *Social Science & Medicine, 65*(8), 1779–1791. https://doi.org/10.1016/j.socscimed.2007.05.051

Jewell, J. O. (2018). 'An injurious effect on the neighbourhood': Narratives of Neighbourhood decline and racialised class identities in late nineteenth-century San Francisco. *Immigrants & Minorities, 36*(1), 1–19. https://doi.org/10.1080/02619288.2017.1355734

Jones, R. W., & Paulson, D. J. (2011). HOPE VI resident displacement: Using HOPE VI program goals to evaluate neighborhood outcomes. *Cityscape, 13*(3), 85–102.

Jones, M., & Pebley, A. R. (2014). Redefining neighborhoods using common destinations: Social characteristics of activity spaces and home census tracts compared. *Demography, 51*(3), 727–752. https://doi.org/10.1007/s13524-014-0283-z

Jung, J.-K., & Elwood, S. (2010). Extending the qualitative capabilities of GIS: Computer-Aided Qualitative GIS. *Transactions in GIS, 14*(1), 63–87. https://doi.org/10.1111/j.1467-9671.2009.01182.x

Keene, D. E., & Geronimus, A. T. (2011). "Weathering" HOPE VI: The importance of evaluating the population health impact of public housing demolition and displacement. *Journal of Urban Health, 88*(3), 417–435. https://doi.org/10.1007/s11524-011-9582-5

Kent-Stoll, P. (2020). The racial and colonial dimensions of gentrification. *Sociology Compass*. https://doi.org/10.1111/soc4.12838

King, K. E., Morenoff, J. D., & House, J. S. (2011). Neighborhood context and social disparities in cumulative biological risk factors. *Psychosomatic Medicine, 73*(7), 572–579. https://doi.org/10.1097/PSY.0b013e318227b062

Krieger, N. (1994). Epidemiology and the web of causation: Has anyone seen the spider? *Social Science & Medicine, 39*(7), 887–903. https://doi.org/10.1016/0277-9536(94)90202-X

Krieger, N. (2001). Theories for social epidemiology in the 21st century: An ecosocial perspective. *International Journal of Epidemiology, 30*(4), 668–677. https://doi.org/10.1093/ije/30.4.668

Krieger, N. (2005). Embodiment: A conceptual glossary for epidemiology. *Journal of Epidemiology & Community Health, 59*(5), 350–355. https://doi.org/10.1136/jech.2004.024562

Kwan, M.-P. (2000). Gender differences in space-time constraints. *Area, 32*(2), 145–156. https://doi.org/10.1111/j.1475-4762.2000.tb00125.x

Kwan, M.-P. (2009). From place-based to people-based exposure measures. *Social Science & Medicine, 69*(9), 1311–1313. https://doi.org/10.1016/j.socscimed.2009.07.013

Kwan, M.-P. (2012). How GIS can help address the uncertain geographic context problem in social science research. *Annals of GIS, 18*(4), 245–255. https://doi.org/10.1080/19475683.2012.727867

Kwan, M.-P., & Ding, G. (2008). Geo-narrative: Extending Geographic information systems for narrative analysis in qualitative and mixed-method research*. *The Professional Geographer, 60*(4), 443–465. https://doi.org/10.1080/00330120802211752

Leung, M., & Takeuchi, D. T. (2011). Race, place, and health. In L. M. Burton, S. A. Matthews, M. Leung, S. P. Kemp, & D. T. Takeuchi (Eds.), *Communities, neighborhoods, and health* (pp. 73–88). Springer. https://doi.org/10.1007/978-1-4419-7482-2_5

Leventhal, T., & Brooks-Gunn, J. (2003). Moving to opportunity: An experimental study of neighborhood effects on mental health. *AJPH, 93*(9), 1576–1582.

Levy, D. K., & Woolley, M. (2007). *Employment barriers among HOPE VI families* (p. 10). The Urban Institute.

Lin, L., & Moudon, A. V. (2010). Objective versus subjective measures of the built environment, which are most effective in capturing associations with walking? *Health & Place, 16*(2), 339–348. https://doi.org/10.1016/j.healthplace.2009.11.002

Ludwig, J. (2011). Neighborhoods, obesity, and diabetes: A randomized social experiment. *New England Journal of Medicine, 365*, 1509–1519.

Macintyre, S., Ellaway, A., & Cummins, S. (2002). Place effects on health: How can we conceptualise and measure them? *Social Science & Medicine, 55*(1), 125–139.

Manjarrez, C. A., Popkin, S. J., & Guernsey, E. (2007). *Poor health: Adding insult to injury for HOPE VI families*. [Data set]. The Urban Institute. https://doi.org/10.1037/e725572011-001

Maryland DHMH. (n.d.). *Maryland Department of Health and Mental Hygiene: Health enterprise zones*. Retrieved February 27, 2013, from http://dhmh.maryland.gov/healthenterprisezones/SitePages/About_Hez.aspx

Matthews, S. A. (2011). Spatial polygamy and the heterogeneity of place: Studying people and place via egocentric methods. In L. M. Burton, S. A. Matthews, M. Leung, S. P. Kemp, & D. T. Takeuchi (Eds.), *Communities, neighborhoods, and health* (pp. 35–55). Springer. https://doi.org/10.1007/978-1-4419-7482-2_3

Matthews, S. A., & Yang, T.-C. (2013). Spatial Polygamy and Contextual Exposures (SPACEs): Promoting activity space approaches in research on place and health. *American Behavioral Scientist, 57*(8), 1057–1081. https://doi.org/10.1177/0002764213487345

Merkin, S. S., Basurto-Dávila, R., Karlamangla, A., Bird, C. E., Lurie, N., Escarce, J., & Seeman, T. (2009). Neighborhoods and cumulative biological risk profiles by race/ethnicity in a national sample of U.S. adults: NHANES III. *Annals of Epidemiology, 19*(3), 194–201. https://doi.org/10.1016/j.annepidem.2008.12.006

Moore, K., Diez Roux, A. V., Auchincloss, A., Evenson, K. R., Kaufman, J., Mujahid, M., & Williams, K. (2013). Home and work neighbourhood environments in relation to body mass index: The Multi-Ethnic Study of Atherosclerosis (MESA). *Journal of Epidemiology and Community Health, 67*(10), 846–853. https://doi.org/10.1136/jech-2013-202682

Morello-Frosch, R., & Lopez, R. (2006). The riskscape and the color line: Examining the role of segregation in environmental health disparities. *Environmental Research, 102*(2), 181–196. https://doi.org/10.1016/j.envres.2006.05.007

Morello-Frosch, R., & Shenassa, E. D. (2006). The environmental "Riskscape" and social inequality: Implications for explaining maternal and child health disparities. *Environmental Health Perspectives, 114*(8), 1150–1153. https://doi.org/10.1289/ehp.8930

Morello-Frosch, R., Pastor, M., & Sadd, J. (2001). Environmental justice and Southern California's "Riskscape": The distribution of air toxics exposures and health risks among diverse communities. *Urban Affairs Review, 36*(4), 551–578. https://doi.org/10.1177/10780870122184993

Moreton-Robinson, A. (2015). *The white possessive: Property, power, and indigenous sovereignty.* University of Minnesota Press. https://www.upress.umn.edu/book-division/books/the-white-possessive

Mujahid, M. S., Diez Roux, A. V., Morenoff, J. D., & Raghunathan, T. (2007). Assessing the measurement properties of neighborhood scales: From psychometrics to ecometrics. *American Journal of Epidemiology, 165*(8), 858–867. https://doi.org/10.1093/aje/kwm040

Nazmi, A., Diez Roux, A., Ranjit, N., Seeman, T. E., & Jenny, N. S. (2010). Cross-sectional and longitudinal associations of neighborhood characteristics with inflammatory markers: Findings from the multi-ethnic study of atherosclerosis☆☆☆. *Health & Place, 16*(6), 1104–1112. https://doi.org/10.1016/j.healthplace.2010.07.001

NCHE. (n.d.). *National collaborative for health equity: Place matters initiative.* Retrieved September 30, 2015, from http://nationalcollaborative.org/?q=node/37

Neely, B., & Samura, M. (2011). Social geographies of race: Connecting race and space. *Ethnic and Racial Studies, 34*(11), 1933–1952. https://doi.org/10.1080/01419870.2011.559262

NHLP. (2002). *False HOPE: A critical assessment of the HOPE VI public housing redevelopment program.* National Housing Law Project.

O'Campo, P. (2003). Invited commentary: Advancing theory and methods for multilevel models of residential neighborhoods and health. *American Journal of Epidemiology, 157*(1), 9–13. https://doi.org/10.1093/aje/kwf171

Osypuk, T. L., & Acevedo-Garcia, D. (2010). Beyond individual neighborhoods: A geography of opportunity perspective for understanding racial/ethnic health disparities. *Health & Place, 16*(6), 1113–1123. https://doi.org/10.1016/j.healthplace.2010.07.002

Pampalon, R., Hamel, D., De Koninck, M., & Disant, M.-J. (2007). Perception of place and health: Differences between neighbourhoods in the Québec City region. *Social Science & Medicine, 65*(1), 95–111. https://doi.org/10.1016/j.socscimed.2007.02.044

Pearce, J. (2015). Invited commentary: History of place, life course, and health inequalities–Historical geographic information systems and epidemiologic research. *American Journal of Epidemiology, 181*(1), 26–29. https://doi.org/10.1093/aje/kwu312

Perchoux, C., Chaix, B., Cummins, S., & Kestens, Y. (2013). Conceptualization and measurement of environmental exposure in epidemiology: Accounting for activity space related to daily mobility. *Health & Place, 21*, 86–93. https://doi.org/10.1016/j.healthplace.2013.01.005

Petersen, K. L., Marsland, A. L., Flory, J., Votruba-Drzal, E., Muldoon, M. F., & Manuck, S. B. (2008). Community socioeconomic status is associated with circulating interleukin-6 and C-reactive protein. *Psychosomatic Medicine, 70*(6), 646–652. https://doi.org/10.1097/PSY.0b013e31817b8ee4

Petteway, R. (2017). Real limits of imaginary lines: A participatory activity space method for exploring intergenerational (Dis)connections between 'place' and health. In *145th meeting of the american public health association.*

Petteway, R. (2018, May 7). The real limits of census tracts, and other boundaries. *Shelterforce.* https://shelterforce.org/2018/05/07/the-real-limits-of-imaginary-lines/

Petteway, R. J. (2019). Intergenerational photovoice perspectives of place and health in public housing: Participatory coding, theming, and mapping in/of the "structure struggle". *Health & Place, 60*, 102229. https://doi.org/10.1016/j.healthplace.2019.102229

Petteway, R. (2021, January 20). Let's re-place the health opportunity maps. *Shelterforce.* https://shelterforce.org/2021/01/20/re-placing-geographies-of-health-opportunity/

Petteway, R., Mujahid, M., & Allen, A. (2019a). Understanding embodiment in place-health research: Approaches, limitations, and opportunities. *Journal of Urban Health, 96*(2), 289–299. https://doi.org/10.1007/s11524-018-00336-y

Petteway, R., Mujahid, M., Allen, A., & Morello-Frosch, R. (2019b). Towards a people's social epidemiology: Envisioning a more inclusive and equitable future for social epi research and practice in the 21st century. *International Journal of Environmental Research and Public Health, 16*(20), 3983. https://doi.org/10.3390/ijerph16203983

Pickett, K. E., & Pearl, M. (2001). Multilevel analyses of neighbourhood socioeconomic context and health outcomes: A critical review. *Journal of Epidemiology & Community Health, 55*(2), 111–122.

Popkin, S. (2004). *A decade of HOPE VI: Research findings and policy challenges.* The Urban Institute.

Popkin, S. J., Levy, D. K., Harris, L. E., Comey, J., Cunningham, M. K., & Buron, L. F. (2004). The HOPE VI program: What about the residents? *Housing Policy Debate, 15*(2), 385–414. https://doi.org/10.1080/10511482.2004.9521506

Powell, J. A., & Bullard, R. (2007). Structural racism and spatial Jim Crow. In *The black metropolis in the twenty-first century: Race, power, and the politics of place* (p. 41).

Powell, J. A., & Cardwell, K. (2013). Homeownership, wealth, & the production of racialized space. *Joint Center for Housing Studies Harvard University*, Article IR. https://lawcat.berkeley.edu/record/1125958

Rainham, D., McDowell, I., Krewski, D., & Sawada, M. (2010). Conceptualizing the healthscape: Contributions of time geography, location technologies and spatial ecology to place and health research. *Social Science & Medicine, 70*(5), 668–676. https://doi.org/10.1016/j.socscimed.2009.10.035

Riva, M., Gauvin, L., & Barnett, T. A. (2007). Toward the next generation of research into small area effects on health: A synthesis of multilevel investigations published since July 1998. *Journal of Epidemiology and Community Health, 61*(10), 853–861. https://doi.org/10.1136/jech.2006.050740

Riva, M., Apparicio, P., Gauvin, L., & Brodeur, J.-M. (2008). Establishing the soundness of administrative spatial units for operationalising the active living potential of residential environments: An exemplar for designing optimal zones. *International Journal of Health Geographics, 7*(1), 43. https://doi.org/10.1186/1476-072X-7-43

Roux, A.-V. D. (2007). Neighborhoods and health: Where are we and were do we go from here? *Revue d'epidemiologie et de Sante Publique, 55*(1), 13–21.

Rudolph, K. E., Gary, S. W., Stuart, E. A., Glass, T. A., Marques, A. H., Duncko, R., & Merikangas, K. R. (2014). The association between cortisol and neighborhood disadvantage in a U.S. population-based sample of adolescents. *Health & Place, 25*, 68–77. https://doi.org/10.1016/j.healthplace.2013.11.001

Ruel, E., Oakley, D., Wilson, G. E., & Maddox, R. (2010). Is public housing the cause of poor health or a safety net for the unhealthy poor? *Journal of Urban Health, 87*(5), 827–838. https://doi.org/10.1007/s11524-010-9484-y

Ruglis, J. (2011). Mapping the biopolitics of school dropout and youth resistance. *International Journal of Qualitative Studies in Education, 24*(5), 627–637. https://doi.org/10.1080/09518398.2011.600268

Sampson, R. J., Morenoff, J. D., & Gannon-Rowley, T. (2002). Assessing "Neighborhood Effects": Social processes and new directions in research. *Annual Review of Sociology, 28*(1), 443–478. https://doi.org/10.1146/annurev.soc.28.110601.141114

Santos, S. M., Chor, D., Werneck, G. L., & Coutinho, E. S. F. (2007). Associação entre fatores contextuais e auto-avaliação de saúde: Uma revisão sistemática de estudos multinível. *Cadernos de Saúde Pública, 23*, 2533–2554.

Schulz, A. J., Mentz, G., Lachance, L., Zenk, S. N., Johnson, J., Stokes, C., & Mandell, R. (2013). Do observed or perceived characteristics of the neighborhood environment mediate associations between neighborhood poverty and cumulative biological risk? *Health & Place, 24*, 147–156. https://doi.org/10.1016/j.healthplace.2013.09.005

Shabazz, R. (2015). *Spatializing blackness: Architectures of confinement and black masculinity in Chicago.* University of Illinois Press.

Shiffman, S., Stone, A. A., & Hufford, M. R. (2008). Ecological momentary assessment. *Annual Review of Clinical Psychology, 4*(1), 1–32. https://doi.org/10.1146/annurev.clinpsy.3.022806.091415

Slater, T. (2013). Your life chances affect where you live: A critique of the 'Cottage Industry' of neighbourhood effects research: A critique of neighbourhood effects research. *International Journal of Urban and Regional Research, 37*(2), 367–387. https://doi.org/10.1111/j.1468-2427.2013.01215.x

Smith, S. J. (2005). The strange geography of health inequalities. *Transactions of the Institute of British Geographers, New Series, 30*(2), 173–190.

Spielman, S. E., & Yoo, E. (2009). The spatial dimensions of neighborhood effects. *Social Science & Medicine, 68*(6), 1098–1105. https://doi.org/10.1016/j.socscimed.2008.12.048

Spook, J. E., Paulussen, T., Kok, G., & Van Empelen, P. (2013). Monitoring dietary intake and physical activity electronically: Feasibility, usability, and ecological validity of a mobile-based ecological momentary assessment tool. *Journal of Medical Internet Research, 15*(9). https://doi.org/10.2196/jmir.2617

Squires, G. D., & Kubrin, C. E. (2005). Privileged places: Race, uneven development and the geography of opportunity in urban America. *Urban Studies, 42*(1), 47–68. https://doi.org/10.1080/0042098042000309694

TCE. (n.d.). *The California endowment: Building healthy communities program.* Retrieved February 27, 2013, from http://www.calendow.org/healthycommunities/

Theall, K. P., Drury, S. S., & Shirtcliff, E. A. (2012). Cumulative neighborhood risk of psychosocial stress and allostatic load in adolescents. *American Journal of Epidemiology, 176*(suppl_7), S164–S174. https://doi.org/10.1093/aje/kws185

Tran, E., Blankenship, K., Whittaker, S., Rosenberg, A., Schlesinger, P., Kershaw, T., & Keene, D. (2020). My neighborhood has a good reputation: Associations between spatial stigma and health. *Health & Place, 64*, 102392. https://doi.org/10.1016/j.healthplace.2020.102392

van Deurzen, I., Rod, N. H., Christensen, U., Hansen, Å. M., Lund, R., & Dich, N. (2016). Neighborhood perceptions and allostatic load: Evidence from Denmark. *Health & Place, 40*, 1–8. https://doi.org/10.1016/j.healthplace.2016.04.010

Van Wart, S., Tsai, K., & Parikh, T. (2010). Local ground: A paper-based toolkit for documenting local geospatial knowledge. *ACM Symposium on Computing for Development (DEV).*

Wang, C. C. (1999). Photovoice: A participatory action research strategy applied to women's health. *Journal of Women's Health, 8*(2), 185–192.

Wang, C., & Burris, M. A. (1997). Photovoice: Concept, methodology, and use for participatory needs assessment. *Health Education & Behavior, 24*(3), 369–387. https://doi.org/10.1177/109019819702400309

Weden, M. M., Carpiano, R. M., & Robert, S. A. (2008). Subjective and objective neighborhood characteristics and adult health. *Social Science & Medicine, 66*(6), 1256–1270. https://doi.org/10.1016/j.socscimed.2007.11.041

Wen, M., Hawkley, L. C., & Cacioppo, J. T. (2006). Objective and perceived neighborhood environment, individual SES and psychosocial factors, and self-rated health: An analysis of older adults in Cook County, Illinois. *Social Science & Medicine, 63*(10), 2575–2590. https://doi.org/10.1016/j.socscimed.2006.06.025

Whitehurst, G. J., & Croft, M. (2010). *The Harlem Children's Zone, Promise neighborhoods, and the broader, bolder approach to education* (p. 12). The Brookings Institute.

Wong, D. W. S., & Shaw, S.-L. (2011). Measuring segregation: An activity space approach. *Journal of Geographical Systems, 13*(2), 127–145. https://doi.org/10.1007/s10109-010-0112-x

Chapter 4
The Real Limits of Imaginary Lines: A Participatory Activity Space Method for Exploring Intergenerational (Dis) Connections Between "Place" and Health

(Mis)Representation

Imaginary Lines + "Senseless Tracts?"

I teach an upper-level undergraduate course on "gender, race, class, and health," which serves mostly as a cursory survey of some of the core "isms" shaping population health. Essentially, we get together twice per week to talk about things like wealth inequality, structural racism, sexism, heterosexism, and so on, and how these forms/processes of oppression and social exclusion affect health opportunities across generations and our lifecourse. And sometimes, we get to do this in a basement classroom without windows in the oldest building on campus. They love it, obviously. As someone with a particular interest in notions of place, I developed a module for the course accordingly: "Place, PlaceMaking, and the People's Health: Pizza Man Don't Come Here No More." It features a selection of readings, documentary films, and music that speak to a range of place and place-making-related concepts, from the Homestead Act of 1862 and Oregon racial exclusion laws, to redlining and racially restrictive covenants, to zoning and gentrification. But before getting into all of this, pointing to a slide containing a polygon-plagued map, I ask the class a simple question: what do these lines (polygons) represent?

© Springer Nature Switzerland AG 2022
R. J. Petteway, *Representation, Re-Presentation, and Resistance*, Global Perspectives on Health Geography, https://doi.org/10.1007/978-3-031-06141-7_4

"ZIP Codes?!!"

"Counties?"

"Neighborhoods?!"

"Congressional districts?"

"Jean Miro's lost cartography thesis?"

Nope. Census tracts.

"Did you say 'senseless' tracts?"

No, but one could argue that they…

As you might imagine, next to no one has ever heard of a census tract by the time they make their ways into my windowless isms dungeon. There are roughly 50 students each term, and each term I offer a 10-point bonus assignment as part of this module: use whatever tech device you have and find out the census tract you lived in when you were 12 years old. For students who did not live in the United States at age 12, I ask them to find something generally comparable. I give them until the end of the class period (roughly 100 min) to submit what they find via our online course platform. Maybe 20 or so students give it a go, with about 4 or 5 of them correctly identifying a/their census tract. College students. In the 2020s. With all the internets out there. And 5G LTE?!

Yes, one could argue…

So what do, and what *can*, these lines represent, then, if most folks don't even know that they exist? And what does it mean that my field of public health, and social epidemiology and place-health research specifically, relies on these lines to construct spatial narratives of place and health—imagining that people actually live *in* them, and fully within them?

As discussed in previous chapters, major conceptual and methodological challenges remain in defining "place," characterizing place contexts, and measuring place (Albright et al., 2011; Bambra et al., 2019; Boruff et al., 2012; Chaix et al., 2009; Cummins, 2007; Cummins et al., 2007; Cutchin et al., 2011; Kwan, 2009; Macintyre et al., 2002; Matthews, 2011; Mujahid et al., 2007; Perchoux et al., 2013; Rainham et al., 2010; Riva et al., 2008; Spielman & Yoo, 2009). While the historic significance and continued importance of census tracts in population health research cannot and should not be discounted (Krieger, 2006, 2019), our overreliance and borderline dependency on these borders impede both empirical and conceptual advancement. A growing body of work makes it clear that the standard practice of defining "place" based on administrative bounds, like census tracts—or worse, ZIP codes—fails to capture the lived spatial reality of peoples' daily place-based health experiences/exposures and opportunities (Crawford et al., 2014; Hand et al., 2018; Inagami et al., 2007; Laatikainen et al., 2018; Matthews & Yang, 2013; Setton et al., 2011; Sharp et al., 2015). Moreover, as I suggest in this and the latter chapters, our affinity for administratively defined constructions of "place" functions to buttress colonial, dispossessing, and disempowering representations and narratives of people's health geographies.

Fortunately, scholarship focusing on place and health continues to grow (Arcaya et al., 2016; Diez Roux & Mair, 2010), including work that articulates the need for spatially dynamic, relational, and boundary-delimited approaches to understanding place-health relationships (Browning & Soller, 2014; Chaix et al., 2009; Cummins, 2007; Cummins et al., 2007; Cutchin et al., 2011; Vallée et al., 2015). One promising approach involves processes to map peoples' daily action or activity spaces—that is, the geographic and social spaces people move within, to, and through as a part of their routines, and the spatiotemporal patterns and interrelations of places therein (Cagney et al., 2020; Perchoux et al., 2013). With roots in space-time and feminist geography (Golledge & Stimson, 1996; Hägerstrand, 1970; Kwan, 2000), *activity space mapping* (ASM) approaches have been used to examine spatial and relational aspects of both youth and adult place-based health exposures, on topics ranging from food environments (Kestens et al., 2010; Raskind et al., 2020; Sadler et al., 2016; Widener et al., 2018; Zenk et al., 2011), tobacco and alcohol environments (Basta et al., 2010; Lipperman-Kreda et al., 2015), greenspace (Bell et al., 2015), and mental health (Vallée et al., 2011), to broader dimensions of social and community context (Jones & Pebley, 2014; Kwan, 2013; Shareck et al., 2014; Wong & Shaw, 2011; York Cornwell & Cagney, 2017).

Even so, to date, there is comparatively limited work that incorporates ASM, even less that simultaneously incorporates both adult *and* youth spaces and perspectives (see, e.g., Browning et al., 2017), and still less that uses a participatory approach. However, information and communication technologies, or ICTs (e.g., smartphones, web-based mapping platforms), present low-cost, readily accessible avenues to support participatory ASM work capable of capturing and integrating place-health experiences/exposures that are both spatially and generationally specific. Accordingly, such work can help distinguish salient places/spaces and exposures from those that are spatially/experientially irrelevant, and thus help elucidate considerations regarding misspecification of place-effects and better guide development of place-based strategies. And, importantly, such work suggests a path to democratize and decolonize the construction/(re)production of place-health narratives—allowing community residents to retain greater narrative control over their place-health geographies.

Here, I present work that supports as much. Drawing from an intergenerational participatory research project examining place, embodiment, and health, this chapter highlights the process and findings for a participatory ASM method and discusses implications for community-inclusive participatory methods in improving conceptual and empirical understanding of how place impacts health. Specifically, this chapter speaks to concerns within public health and health geography regarding misspecification of place-health effects and misrepresentation of place-health experiences, presenting the participatory activity space approach as a way to improve our collective work going forward by: (a) accounting for peoples' daily mobility patterns in defining "place" and place-health exposures, (b) including intergenerational perspectives of place-health contexts, and (c) making use of ICTs to facilitate a more inclusive and participatory research practice that can reveal important socio-spatial (dis)connections that shape place-health experiences.

Below, I detail the ASM process and summarize core findings. I then discuss the potential importance of participatory ASM approaches not only for improving empirical work on place-health relationships but also—and more importantly—for improving conceptual understanding of place. In doing so, I center considerations of power and agency within community assessment and mapping practices, suggesting a need for greater awareness of how such practices can function as exclusionary placemaking mechanisms of spatial stigma and misrepresentation, while simultaneously obscuring critical sociospatial contexts that shape spatial behaviors.

Re-Presentation

Participatory Activity Space Mapping of Place-Health Geographies

Background

As part of an intergenerational community-based participatory research project examining place and health—the *People's Social Epi Project*, or PSEP (as described in CH3)—participants were recruited as parent-youth dyads and trained in four participatory methods, including photovoice (via smartphones), activity space mapping (ASM), X-Ray Mapping, and web-based participatory GIS. They used this combination of methods in succession—first photovoice, then ASM, then X-Ray Mapping, then PGIS—to document and geolocate their daily place-based health experiences/exposures. Participants used large printed maps, worksheets, and a multimedia-enabled web-based mapping platform to complete ASM and digitally represent and illustrate their activity spaces, with two key outputs in addition to maps: *PlaceTime* and *PlaceGrades*. The following sections detail the process and findings specifically for the ASM approach.

Participatory Activity Space Mapping Process + Methods

Activity Space Mapping

For ASM, participants used large print-out maps to identify the locations of their photovoice photos using stickers, markers, and note tabs (Photo 4.1). Participants then identified additional important places for which they had not taken photos. This process was completed over the course of three meetings. Participants were instructed to use green stickers to represent photo-places that were healthy/good/positive, red stickers for photo-places that were unhealthy/bad/negative, and yellow stickers for photo-places that they perceived as both healthy and unhealthy (Photo 4.2), using note tabs to label places as desired (Photo 4.3).

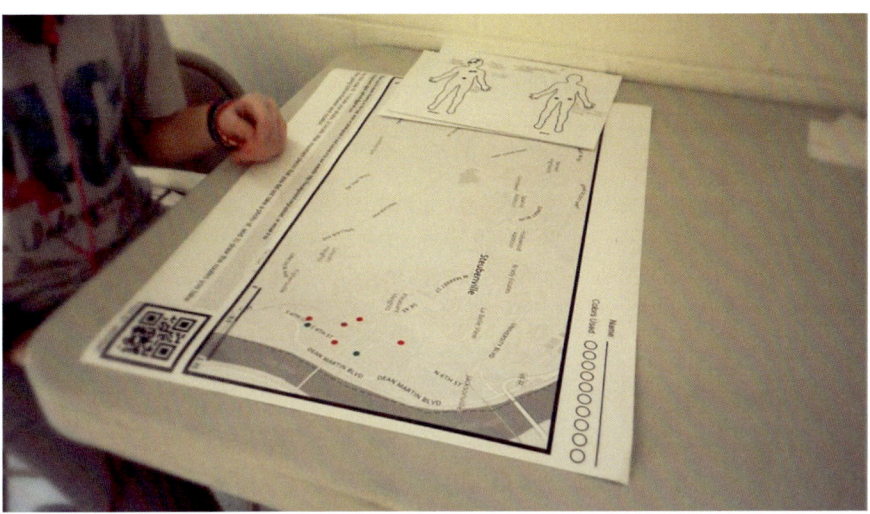

Photo 4.1 Youth participant completing a printed Activity Space Map. Photo of a youth participant using color-coded stickers to spatially locate important place-health experiences. They used
printed maps from a web-based community mapping platform, *Local Ground*, for this part of the
process. Green stickers represented "healthy/positive/good" places; red represented "unhealthy/
negative/bad" place; yellow represented places that were both healthy and unhealthy

Photo 4.2 Example printed Activity Space Mapping map, #1. Photo of a completed activity space
map. Green stickers represented "healthy/positive/good" places; red represented "unhealthy/nega-
tive/bad" place; yellow represented places that were both healthy and unhealthy

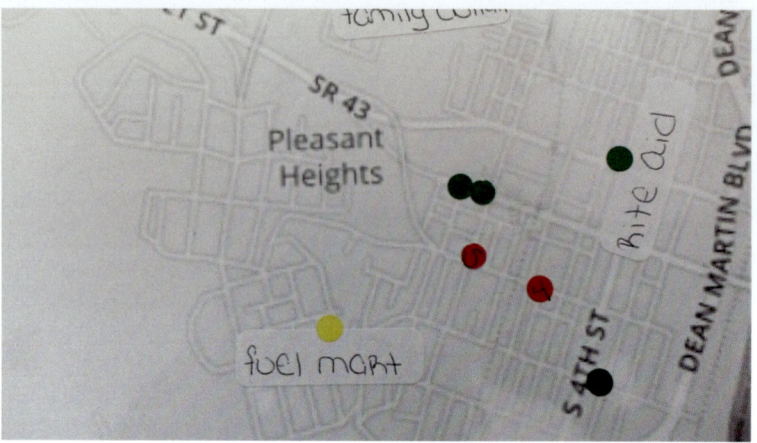

Photo 4.3 Example printed Activity Space Mapping Map, #2. Photo of a completed activity space map that includes location labels. Participants were encouraged to label their place-health locations however they chose, and to also draw travel routes taken between places. Green stickers represented "healthy/positive/good" places; red represented "unhealthy/negative/bad" place; yellow represented places that were both healthy and unhealthy

Participants also completed an ASM worksheet for each mapped place (Fig. 4.1). This worksheet asked them to estimate how much time they spend in each place, how long it takes to travel to and from, how they travel (e.g., walk, bus), and how often they go there. Participants also expressed a desire to "rate their place" and decided to use a star scale to do so, with 1-star being the lowest (e.g., very unhealthy) and 5-stars being the highest (e.g., very healthy) (Fig. 4.2). These data were aggregated for adults and youth separately and used to develop two metrics decided upon by the participants: *PlaceTime* and *PlaceGrade*. *PlaceTime* is based on the estimates participants made for the amount of time (minutes per day) spent at each place identified via their ASM worksheets. *PlaceGrade* is based on the star-rating participants assigned to each place identified via their ASM worksheets—with the five star-levels transformed into a five-point grade point average scale with 12 categories, where 1 = F and 5 = A. So, for example, if the average star rating a participant gave to their places in the *Neighborhood* place-domain was 2.5 stars, their place "GPA" would be 2.5—and thus a *Neighborhood PlaceGrade* of D+.

Participatory GIS

For *Participatory GIS*, participants synthesized and uploaded their data (photos, narratives, positive/negative body-effects, etc.) to the web-based *Local Ground* platform (Van Wart et al., 2010). This platform allowed participants to create a password-protected account (similar to an email account) to upload and digitally map their data. The printed ASM maps had QR codes for scanning and digital translation of the printed maps; however, not all participants were interested in using this function.

PEOPLE'S SOCIAL EPI PROJECT: ACTIVITY SPACE MAPPING GUIDE

Your Name: _____

Photo # and title: _____

Photo Location/Address: _____

For each photo you've mapped out, write/type a brief response to each of the following.

1) How do you usually get here (for example, walk, drive, bus, dropped off)?

2) How long does it take to get here (in minutes)?

3) How much time do you usually spend when you're here (in minutes)?

4) When do you go here (for example, which days of the week, what time of day, how often, before/after what)?

5) Why do you go here?

6) What do you do here?

7) Would you say this place is important to you or your community? Why or why not?

8) Is there anything that you feel is particularly healthy or unhealthy about this place? Or anything you feel is particularly good or bad about it?

RATE YOUR PLACE

If you could rate how healthy/good for your health or your community's health this place is, or how unhealthy/bad for your health or your community's this place is, how would you rate it?

Give your photo-place a rating between 1 and 5 Stars.
1 Star = Very *unhealthy*, bad, or negative
5 Stars = Very healthy, good, or positive

☆☆☆☆☆

Other comments or thoughts:

Fig. 4.1 Activity Space Mapping guide. ASM worksheet. Participants completed one of these worksheets for each place-health location they mapped out using the printed maps

Place-Domain Categorization

Participants' data elements were assigned to a place-domain based on the data topic and location. For example, a photo of a participant's housing environment would be assigned to the "Home" domain, a photo related to a participant's school/place of work would be assigned to the "School/Work" domain, and so on. Data reflecting

RATE YOUR PLACE

If you could rate how healthy/good for your health or your community's health this place is, or how unhealthy/bad for your health or your community's this place is, how would you rate it?

Give your photo-place a rating between 1 and 5 Stars.
1 Star = Very *unhealthy*, bad, or negative
5 Stars = Very healthy, good, or positive

☆☆☆☆☆

Other comments or thoughts:

Fig. 4.2 "Rate Your Place" section of Activity Space Mapping guide. "Rate Your Place" section of the ASM worksheets participants completed. Inclusion of this section was at the suggestion of youth co-researchers who thought it would be cool to create a sort of place-health "Yelp." Adding this section required submitting an Institutional Review Board amendment that delayed youths' work for about 2 months

their community built, social, and food environments, etc. were assigned to the "Neighborhood" domain, except those data for which associated narratives indicated that a particular location was simply observed/passed on their route/way to another intended destination (e.g., "I walk by this building on the way to school"). In this case, data were assigned to the "Transition" domain *and* "Neighborhood" domain, but counted only in the "Transition" domain for the data presented here. Data related to leisure/social activities or related places were assigned to the "Leisure/Social" domain.

Participatory ASM Findings

Youth completed a total of 43 activity space maps, and adults completed a total of 21. Tables 4.1 and 4.2 summarize youth and adult ASM data across the five place-domains. Data for all participants were aggregated and averaged for each place-domain. So, for example, based on the data they provided via their Activity Space Maps, youth spent an average of 887 min per day (*PlaceTime*) in places within their "Home" place-domain (e.g., in their own housing unit, in the building hallways, in common spaces). The average star-rating they assigned to these Home place-domain places was 1.6 stars (*PlaceScore*), which translates to a letter grade of "F" on a 5.0 grading scale (*PlaceGrade*). For the "Transition" place-domain, tabulations for *PlaceTime, PlaceScore*, and *PlaceGrade* were made only for data corresponding to Activity Space Maps that were specifically related to their transition routes, i.e., those explicitly evaluating aspects of their travel routes. Thus an Activity Space Map related to a "Leisure/Social" place, for example, might contain data on travel time to/from that place (e.g., 12 min), but the focus of that activity space map and remaining data is the intended *destination*, not the journey. These activity space maps thus contain ungraded transition times. These ungraded transition times are shown in light-orange, e.g., 17 min per day transitioning to/from school among youth participants, for which they did not "grade" the healthy/unhealthiness of their transition experience/contexts.

Table 4.1 Youth Activity Space Mapping summary

	Avg. PlaceTime (min./day)	Avg. PlaceScore (1 to 5)	PlaceGrade
Home	887	1.6	F
Transition (Home)	3		
Neighborhood	9	2.34	D
Transition (Neighborhood)	8		
School	420	3	C
Transition (School)	17		
Leisure/Social	72	4.75	A-
Transition (Leisure/Social)	12		
Transition	14	2.42	D+
Total PlaceTime	1440	2.82	C-

Summary of youth participants' ASM data based on their completed ASM worksheets. The "Rate Your Place" section of the ASM worksheet (described above) was converted to a numeric rating initially, *PlaceScore*. Youth co-researchers then suggested converting it to a letter-grade rating, *PlaceGrade*

Table 4.2 Adult Activity Space Mapping summary

	Avg. PlaceTime (min./day)	Avg. PlaceScore (1 to 5)	PlaceGrade
Home	1026	1.33	F
Transition (Home)	-		
Neighborhood	11	1.64	F
Transition (Neighborhood)	14		
Work/Errand	263	4	B
Transition (School)	13		
Leisure/Social	90	3.67	C+
Transition (Leisure/Social)	14		
Transition	9	1	F
Total PlaceTime	1440	2.33	D+

Summary of adult participants' ASM data based on their completed ASM worksheets

Figures 4.3 and 4.4 geographically illustrate the positive/negative distribution of place-locations identified by youth and adult participants during photovoice and activity space mapping. The polygon outline is the residential census tract for participants' housing project community, here represented by a single black marker. Green markers represent places participants identified as positive/healthy/good, while red markers represent places identified as negative/unhealthy/bad. Tables 4.3 and 4.4 represent this data in tabular form in relation to participants' census tract of residence. Overall, 55% of adult and 51% of youth places were located spatially outside of their residential census tract. For youth (Table 4.3), 80% of their positive/healthy/good places were outside of their census tract, while 67% of their negative/unhealthy/bad places were inside. Among adults, 62% of positive/healthy/good places were outside of their census tract, while negative/unhealthy/bad places were fairly evenly distributed inside and outside.

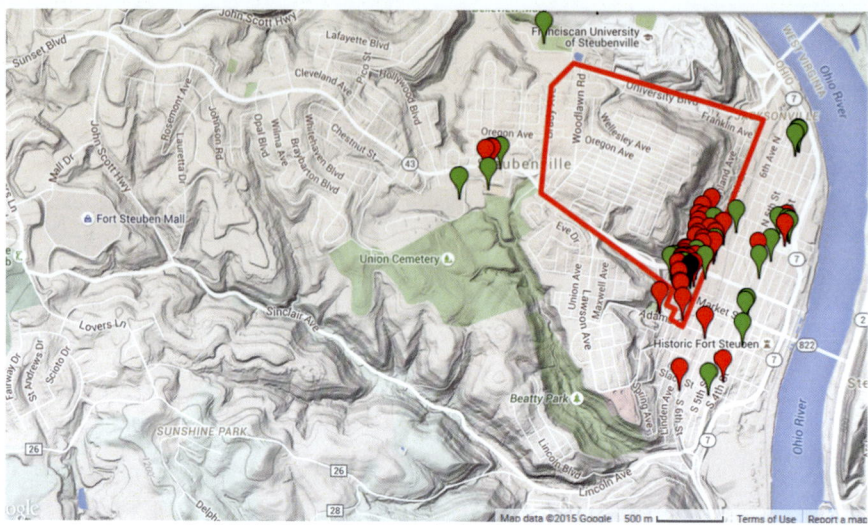

Fig. 4.3 Spatial distribution of youth photovoice and activity space mapping places. Green = positive/healthy/good place. Red = negative/unhealthy/bad place. Black marker = participants' housing community. Red polygon outline = participants' residential census tract. NOTE: the use of red here was due to what appeared to be a technological glitch within the platform coding, which kept returning all polygon color changes back to red

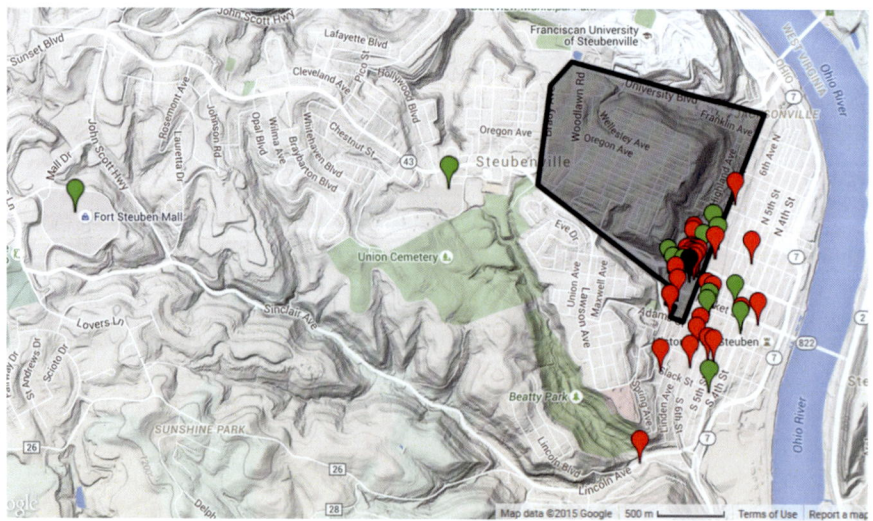

Fig. 4.4 Spatial distribution of adult photovoice and activity space mapping places. Green = positive/healthy/good place. Red = negative/unhealthy/bad place. Black marker = participants' housing community. Black polygon outline = participants' residential census tract

Table 4.3 Youth appraisal of photovoice and activity space mapping places

Youth Photovoice and Activity Space Mapping Places			
	In CT	Outside CT	Farthest Place (miles)
Positive	5 (20%)	20 (80%)	1.7
Negative	28 (67%)	14 (33%)	1.4
Total	33 (49%)	34 (51%)	--

Summary of how youth appraised their place-health locations identified via photovoice and ASM. *CT* census tract. The "Positive" column captures "healthy/positive/good" places as identified during the project. The "Negative" column captures "unhealthy/negative/bad" place as identified during the project

Table 4.4 Adult appraisal of photovoice and activity space mapping places

Adult Photovoice and Activity Space Mapping Places			
	In CT	Outside CT	Farthest Place (miles)
Positive	5 (38%)	8 (62%)	4.3
Negative	14 (48%)	15 (52%)	1.1
Total	19 (45%)	23 (55%)	--

Summary of how adults appraised their place-health locations identified via photovoice and ASM. *CT* census tract. The "Positive" column captures "healthy/positive/good" places as identified during the project. The "Negative" column captures "unhealthy/negative/bad" place as identified during the project

Resistance

Discussion: Participatory Activity Space Geographies as Lines of Resistance

The overall goal in presenting the work discussed in this chapter was to highlight the potential value of fully participatory ASM approaches in improving conceptual and empirical understanding of place-health relationships. In doing, I also sought to engage core considerations of power and agency within traditional spatial and narrative representation processes of place-health geographies, making use of ICTs to center community voice/experience and democratize local place-health knowledge practice. In regard to these matters, the research presented in this chapter suggests perhaps five major points for consideration.

First, the participatory nature of this project allowed participants to retain control in defining the spatial contours of their daily place-health geographies. As a result, and as discussed in Chap. 3, Figs. 4.3 and 4.4 here make it clear that, spatially speaking, the overwhelming majority of participants' residential census tract had no bearing on their place-health experiences. Moreover, over half of the adult and youth place-based health exposures/experiences reported were outside of their residential census tract. Indeed, among youth, 74% of positive place-based exposures/experiences were outside of their census tract. In this case, standard research and community health assessment practices that default the census tract as the "neighborhood" exposure area would have accordingly overestimated potentially negative exposures while simultaneously underestimating potentially positive exposures. And as shown in Tables 4.1 and 4.2, participants spent a significant amount of time outside of their residential census tract. While these *PlaceTime* data are admittedly crude and cursory at best, they do suggest a potential salubrious impact of spending time outside of their immediate neighborhood. Overall, these findings lend further support to literature raising concerns over the misspecification of place-effects (Inagami et al., 2007; Kwan, 2009, 2012; Robinson & Oreskovic, 2013; Roux, 2007; Spielman & Yoo, 2009) and calling for more nuanced approaches to studying place and health (Browning & Soller, 2014; Cummins et al., 2007; Cutchin et al., 2011; Macintyre et al., 2002). Unfortunately, there remains only limited work that includes non-residential places within the aggregate place-health exposure picture (see, e.g., Hoehner et al., 2013; Inagami et al., 2007; Moore et al., 2013), much of which still relies on administratively defined geographies (e.g., census tract of home and census tract of workplace), and none of which has been participatory.

Second, and relatedly, participants' ASM data reflect rather poignantly the significance of Macintyre et al.' (2002) notions of "opportunity structures" and "needs-driven place." As noted, over half of participants' important daily places were located outside of their residential census tract. By and large, with the exception of social gatherings and leisure activities at the community recreation center adjacent to their housing community, existing opportunity structures—and consequent routine activities to meet needs—defined a "place" that was mostly outside of their residential census tract. For example, their schools, grocery stores, retail shops, pharmacies, places of work, and places of worship were all beyond the bounds of their administratively defined neighborhood (though as shown in Chap. 3, some retail shops were literally across the street in a different census tract; so the standard spatial buffer applied in much work would have captured this in some capacity). This was true for both participants' more material needs, as well as social needs. As one prominent example, adults' walking route/loop—a mile away in an entirely different neighborhood and different census tract—in many ways speaks to the work of Williams and Hipp (2019) regarding the importance of accounting for the role—and spatiality—of "third places" within place-health research. While traditional individualist and behaviorist public health might only be interested in the walking itself, e.g., "moderate physical activity," it may be best appreciated from a relational and spatial polygamy perspective in that the walking—as a social practice—represents an act *and* site of placemaking and bonding, with the walking route itself a spatial convergence of multiple activity spaces. In other words, it's not about burning calories, but *being in community*—even if being in community literally means being outside of one's own neighborhood. Standard place-health research practices would have missed this important aspect of participants' place-health geographies entirely, and non-participatory activity space approaches would have ignored its significance as a sociospatial placemaking practice of community, i.e., the "why" behind the "where."

Third, as shown in Figs. 4.3 and 4.4, there were large spatial areas that participants did not appear to go to/through. The Figures might suggest a significant role for topography in shaping participants' spatial behaviors and daily activity spaces. Indeed, this was generally true for most participants, as they did not go "up on the hill" unless truly necessary—namely, to go to school or the grocery store, both of which were outside of their residential census tract. This of course speaks again to the salience of opportunity structures and needs-driven place. Here, in the context of activity space, school represents a "fixed activity" in a fixed place and the grocery store represents a "flexible activity" in a fixed place, and both represent a "coupling" constraint on participants' routine spatial behaviors, i.e., they have to go these places regularly and within set time windows (Hägerstrand, 1970; Perchoux et al., 2013). The grocery store and school required participants to traverse through/ to spaces they seemingly otherwise would not have, i.e., why climb "the hill" if you don't have to? It's certainly important to note that topographic considerations are often entirely omitted within administratively defined assessments of place and health (see, e.g., "walk scores" that ignore elevation gain), and while these data allow for these considerations here, they may actually obscure another important spatial barrier: sociospatial practices and significations that serve to spatially exclude and disconnect people from places.

As noted in Cummins et al. (2007) notion of *relational place*, place is best understood as a dynamic social and cultural production inextricably linked to political and economic processes. This means we need to account for how people perceive and relate to their daily social and physical environments, part of which entails understanding the social and political nature of space and how it informs people's place-based activities. To the extent that people lead spatially polygamous lives with multiple overlapping spatial affinities and places of attachment and belonging (Leung & Takeuchi, 2011; Matthews, 2011), they also lead lives of spatial exclusion and division—often literal and symbolic spatial delineations and significations that keep people "in their place" by keeping them out of spaces (Bonds & Inwood, 2016; Inwood & Yarbrough, 2010; Kwan, 2013; Neely & Samura, 2011; Peake & Ray, 2001; Powell & Cardwell, 2013; Pulido, 2000; Wong & Shaw, 2011). And this, as much as topography, may explain the empty space in Figs. 4.3 and 4.4. All, but two participants in this project were Black. The residential area just at the edge of "the hill" was and historically has been almost exclusively White. There was only one walking route up (via "the steps"), which ended next to a household suspected of routinely calling the police on folks passing through the neighborhood. The sentiment was that participants, as mostly young Black teenagers, did not feel welcomed or safe passing through. It was, in other words, felt to be a "White space" (Anderson, 2015), one which functioned as an "authority" space-time constraint for the participants (Hägerstrand, 1970). Thus, to some extent, the empty part of the map was "off limits" due to racialized spatial surveillance and exclusion. Existing opportunity structures were such that youth and adults had to traverse through spaces of surveillance and exclusion on a routine basis, or, were dissuaded from doing so. In both scenarios, their activity spaces were defined not just by the geographic locations they needed to go in relation to home, but also by sociospatial relations of power that may have animated decisions regarding how (e.g., route, walking, bus) and when to go (e.g., daytime vs. nighttime), if at all. Non-participatory activity space approaches employed to date have failed to engage this dynamic, ignoring important aspects of sociospatial context that shape place-health geographies.

Fourth, generally speaking, participants' work makes it clear why intergenerational approaches to ASM are a valuable consideration within place-health research. As noted in Chaps. 1 and 3, adults and youth have fundamentally different place-health exposures and spatial behaviors. Efforts to move toward more conceptually nuanced and spatially dynamic investigations of place-health relationships largely will be for naught if such approaches fail to account for the implications of life-course epidemiology and the salience/role of developmental age and life-stage in shaping spatial behaviors. A growing body of work accordingly suggests the importance of examining daily mobility and activity spaces in a manner that accounts for residents' age or life stage (Franke et al., 2017; Hirsch et al., 2014; Milton et al., 2015; Villanueva et al., 2012; York Cornwell & Cagney, 2017). While crude, Tables 4.1 and 4.2 do well to illustrate this in regards to the types of places adult and youth participants spent their time, as well as how they appraised them. And Figs. 4.3 and 4.4, along with Tables 4.3 and 4.4, reveal spatial and perceptual differences between adult and youth participants' routine places—perhaps most notably in regards to the distribution of youth's positive/healthy places in relation to their residential census track (80% outside), compared to that of adults (62%). Activity space approaches, in general, allow for more specific and dynamic assessments of place-health exposures. Further refinement to account for age/life-stage considerations enhances their conceptual, empirical, and pragmatic value (e.g., tailoring potential interventions). And here, I suggest, participatory intergenerational ASM approaches enable an even more thorough accounting of place-health exposures based on their perceived significance as encountered, documented, and described by residents themselves. As noted above, much relevant sociospatial context is occluded from spatial narratives and representations otherwise.

Fifth, and in sum, participants' work illustrates the value of anchoring place-health assessment processes in residents' lived experience of place as actually encountered—their place-health geographies were not arbitrarily bounded and analyzed based on an imaginary line that none of them knew existed. This was greatly facilitated by the use of ICTs, with residents being able to use their smartphones and web-based mapping to document and narrate their place-health geographies themselves via a mix of qualitative processes and participatory GIS, versus traditional activity space approaches that rely largely upon use of GPS tracking devices and mostly quantitative metrics to capture/define place (e.g., standard deviation ellipses, kernel density estimation, convex envelope, spatial buffers). The ICT-facilitated process used here allowed for residents to be co-researchers generating data and leading analyses, not simply study "subjects" whose spatial behaviors are technologically surveilled, stripped of context, and anonymously analyzed by "researchers" elsewhere. The former does much better to ensure residents retain agency and narrative power throughout the place-health knowledge production and representation process. This is especially important to note given that the most recent review of the conceptual and empirical promise of activity space approaches for place-health research failed to critically engage considerations of power, (mis)representation, and participation in the knowledge production process (Cagney et al., 2020). Indeed, the words "participatory," "collaboration," "collaborative," "inclusive," "inclusion," and "power" were not mentioned a single time. This suggests a presumptive colonial, extractivist, and elitist orientation as the default, and that such epistemes might continue to dominate this line of research without due interrogation. The use of ICTs in the work presented here not only greatly facilitated alignment with the PSEP project's animating theoretical orientations regarding power and centering margins (see Chap. 3; Petteway et al., 2019), but it also allowed for a more inclusive and democratic representation/re-presentation of residents' geographies of place and health—as counternarrative and resistance to externally generated place-health narratives (see also Chap. 5).

Overall, participants' work—as described here and in Chap. 3—does well to illustrate that "place" is more than just an unchanging, administratively defined geographic location. Rather, it is a dynamic social and cultural production inextricably linked to political and economic processes that unfold on various spatial scales (e.g., city, neighborhood). This means we need to account for how people experience and relate to their daily social and material environments and identify the political and economic forces that not only shape those environments but make it such that it is those specific environments—spatially and sociomaterially—that they are exposed to. As articulated by Neely and Samura (2011) in discussing Lefebvre's analysis of spatiality and political economy, "relations of power intersect with and may force the movement or placement of people; they also inform the knowledge produced within and about particular spaces" (p. 1936). In sum, it's not only that we must improve how we conceptualize, define, and measure place, but also that we must better explicate how places—and specific communities' place contexts—come about, and elucidate sociospatial (dis)connections and divisions therein. Moreover, we must acknowledge that the knowledge we produce about "place"—based on how we choose to (mis)represent it and narrate it—is a political act and exercise of both productive and repressive power (discussed in greater detail in Chap. 6).

Conclusion

In general, the work presented here suggests participatory ASM approaches that are anchored in principles of CBPR, and incorporate both adult and youth perspectives and places, can greatly enhance understanding of how, when, and where place matters for health across generations. Moreover, participants' work highlighted the potential value of ICTs in amplifying community agency and voice in place-health research/practice efforts. ICTs open the door for low-cost "crowdsourcing" approaches for ASM within place-health research and can offer residents opportunities to inform local practice (discussed in detail in Chap. 5). ICTs, with appropriate conceptual and theoretical groundings, afford the opportunity to move toward more inclusive and equitable participatory ASM approaches, which present an avenue to democratize place-health research—to "socialize" the production of place-health knowledge and the narration of place-health geographies.

References

Albright, K., Chung, G., De Marco, A., & Yoo, J. (2011). Moving beyond geography: Health practices and outcomes across time and place. In L. M. Burton, S. A. Matthews, M. Leung, S. P. Kemp, & D. T. Takeuchi (Eds.), *Communities, neighborhoods, and health* (pp. 127–143). Springer. https://doi.org/10.1007/978-1-4419-7482-2_8

Anderson, E. (2015). The white space. *Sociology of Race and Ethnicity, 1*(1), 10–21. https://doi.org/10.1177/2332649214561306

Arcaya, M. C., Tucker-Seeley, R. D., Kim, R., Schnake-Mahl, A., So, M., & Subramanian, S. V. (2016). Research on neighborhood effects on health in the United States: A systematic review of study characteristics. *Social Science & Medicine, 168*, 16–29. https://doi.org/10.1016/j.socscimed.2016.08.047

Bambra, C., Smith, K. E., & Pearce, J. (2019). Scaling up: The politics of health and place. *Social Science & Medicine, 232*, 36–42. https://doi.org/10.1016/j.socscimed.2019.04.036

Basta, L. A., Richmond, T. S., & Wiebe, D. J. (2010). Neighborhoods, daily activities, and measuring health risks experienced in urban environments. *Social Science & Medicine, 71*(11), 1943–1950. https://doi.org/10.1016/j.socscimed.2010.09.008

Bell, S. L., Phoenix, C., Lovell, R., & Wheeler, B. W. (2015). Using GPS and geo-narratives: A methodological approach for understanding and situating everyday green space encounters: Using GPS and geo-narratives. *Area, 47*(1), 88–96. https://doi.org/10.1111/area.12152

Bonds, A., & Inwood, J. (2016). Beyond white privilege: Geographies of white supremacy and settler colonialism. *Progress in Human Geography, 40*(6), 715–733. https://doi.org/10.1177/0309132515613166

Boruff, B. J., Nathan, A., & Nijënstein, S. (2012). Using GPS technology to (re)-examine operational definitions of 'neighbourhood' in place-based health research. *International Journal of Health Geographics, 11*(1), 22. https://doi.org/10.1186/1476-072X-11-22

Browning, C. R., & Soller, B. (2014). Moving beyond neighborhood: Activity spaces and ecological networks as contexts for youth development. *Cityscape (Washington, DC), 16*(1), 165.

Browning, C. R., Calder, C. A., Soller, B., Jackson, A. L., & Dirlam, J. (2017). Ecological networks and neighborhood social organization. *American Journal of Sociology, 122*(6), 1939–1988. https://doi.org/10.1086/691261

Cagney, K. A., York Cornwell, E., Goldman, A. W., & Cai, L. (2020). Urban mobility and activity space. *Annual Review of Sociology, 46*(1), 623–648. https://doi.org/10.1146/annurev-soc-121919-054848

Chaix, B., Merlo, J., Evans, D., Leal, C., & Havard, S. (2009). Neighbourhoods in eco-epidemiologic research: Delimiting personal exposure areas. A response to Riva, Gauvin, Apparicio and Brodeur. *Social Science & Medicine, 69*(9), 1306–1310. https://doi.org/10.1016/j.socscimed.2009.07.018

Crawford, T. W., Jilcott Pitts, S. B., McGuirt, J. T., Keyserling, T. C., & Ammerman, A. S. (2014). Conceptualizing and comparing neighborhood and activity space measures for food environment research. *Health & Place, 30*, 215–225. https://doi.org/10.1016/j.healthplace.2014.09.007

Cummins, S. (2007). Commentary: Investigating neighbourhood effects on health – Avoiding the "Local Trap". *International Journal of Epidemiology, 36*(2), 355–357. https://doi.org/10.1093/ije/dym033

Cummins, S., Curtis, S., Diez-Roux, A. V., & Macintyre, S. (2007). Understanding and representing 'place' in health research: A relational approach. *Social Science & Medicine, 65*(9), 1825–1838. https://doi.org/10.1016/j.socscimed.2007.05.036

Cutchin, M. P., Eschbach, K., Mair, C. A., Ju, H., & Goodwin, J. S. (2011). The socio-spatial neighborhood estimation method: An approach to operationalizing the neighborhood concept. *Health & Place, 17*(5), 1113–1121. https://doi.org/10.1016/j.healthplace.2011.05.011

Diez Roux, A. V., & Mair, C. (2010). Neighborhoods and health: Neighborhoods and health. *Annals of the New York Academy of Sciences, 1186*(1), 125–145. https://doi.org/10.1111/j.1749-6632.2009.05333.x

Franke, T., Winters, M., McKay, H., Chaudhury, H., & Sims-Gould, J. (2017). A grounded visualization approach to explore sociospatial and temporal complexities of older adults' mobility. *Social Science & Medicine, 193*, 59–69. https://doi.org/10.1016/j.socscimed.2017.09.047

Golledge, R., & Stimson, R. (1996). *Spatial behavior: A geographic perspective*. Guilford Press. https://www.guilford.com/books/Spatial-Behavior/Golledge-Stimson/9781572300507

Hägerstrand, T. (1970). What about people in Regional Science? *Papers of the Regional Science Association, 24*(1), 6–21. https://doi.org/10.1007/BF01936872

Hand, C. L., Rudman, D. L., Huot, S., Gilliland, J. A., & Pack, R. L. (2018). Toward understanding person–place transactions in neighborhoods: A qualitative-participatory geospatial approach. *The Gerontologist, 58*(1), 89–100. https://doi.org/10.1093/geront/gnx064

Hirsch, J. A., Winters, M., Clarke, P., & McKay, H. (2014). Generating GPS activity spaces that shed light upon the mobility habits of older adults: A descriptive analysis. *International Journal of Health Geographics, 13*(1), 51. https://doi.org/10.1186/1476-072X-13-51

Hoehner, C. M., Allen, P., Barlow, C. E., Marx, C. M., Brownson, R. C., & Schootman, M. (2013). Understanding the independent and joint associations of the home and workplace built environments on cardiorespiratory fitness and body mass index. *American Journal of Epidemiology, 178*(7), 1094–1105. https://doi.org/10.1093/aje/kwt111

Inagami, S., Cohen, D. A., & Finch, B. K. (2007). Non-residential neighborhood exposures suppress neighborhood effects on self-rated health. *Social Science & Medicine, 65*(8), 1779–1791. https://doi.org/10.1016/j.socscimed.2007.05.051

Inwood, J. F., & Yarbrough, R. A. (2010). Racialized places, racialized bodies: The impact of racialization on individual and place identities. *GeoJournal, 75*(3), 299–301. https://doi.org/10.1007/s10708-009-9308-3

Jones, M., & Pebley, A. R. (2014). Redefining neighborhoods using common destinations: Social characteristics of activity spaces and home census tracts compared. *Demography, 51*(3), 727–752. https://doi.org/10.1007/s13524-014-0283-z

Kestens, Y., Lebel, A., Daniel, M., Thériault, M., & Pampalon, R. (2010). Using experienced activity spaces to measure foodscape exposure. *Health & Place, 16*(6), 1094–1103. https://doi.org/10.1016/j.healthplace.2010.06.016

Krieger, N. (2006). A century of census tracts: Health & the body politic (1906–2006). *Journal of Urban Health, 83*(3), 355–361. https://doi.org/10.1007/s11524-006-9040-y

Krieger, N. (2019). The US census and the people's health: Public health engagement from enslavement and "indians not taxed" to census tracts and health equity (1790–2018). *American Journal of Public Health, 109*(8), 1092–1100. https://doi.org/10.2105/AJPH.2019.305017

Kwan, M.-P. (2000). Gender differences in space-time constraints. *Area, 32*(2), 145–156. https://doi.org/10.1111/j.1475-4762.2000.tb00125.x

Kwan, M.-P. (2009). From place-based to people-based exposure measures. *Social Science & Medicine, 69*(9), 1311–1313. https://doi.org/10.1016/j.socscimed.2009.07.013

Kwan, M.-P. (2012). How GIS can help address the uncertain geographic context problem in social science research. *Annals of GIS, 18*(4), 245–255. https://doi.org/10.1080/19475683.2012.727867

Kwan, M.-P. (2013). Beyond space (as we knew it): Toward temporally integrated geographies of segregation, health, and accessibility: Space–time integration in geography and GIScience.

Annals of the Association of American Geographers, 103(5), 1078–1086. https://doi.org/1 0.1080/00045608.2013.792177

Laatikainen, T. E., Hasanzadeh, K., & Kyttä, M. (2018). Capturing exposure in environmental health research: Challenges and opportunities of different activity space models. *International Journal of Health Geographics, 17*(1), 29. https://doi.org/10.1186/s12942-018-0149-5

Leung, M., & Takeuchi, D. T. (2011). Race, place, and health. In L. M. Burton, S. A. Matthews, M. Leung, S. P. Kemp, & D. T. Takeuchi (Eds.), *Communities, neighborhoods, and health* (pp. 73–88). Springer. https://doi.org/10.1007/978-1-4419-7482-2_5

Lipperman-Kreda, S., Morrison, C., Grube, J. W., & Gaidus, A. (2015). Youth activity spaces and daily exposure to tobacco outlets. *Health & Place, 34*, 30–33. https://doi.org/10.1016/j.healthplace.2015.03.013

Macintyre, S., Ellaway, A., & Cummins, S. (2002). Place effects on health: How can we conceptualise, operationalise and measure them? *Social Science & Medicine, 55*(1), 125–139.

Matthews, S. A. (2011). Spatial polygamy and the heterogeneity of place: Studying people and place via egocentric methods. In L. M. Burton, S. A. Matthews, M. Leung, S. P. Kemp, & D. T. Takeuchi (Eds.), *Communities, neighborhoods, and health* (pp. 35–55). Springer. https://doi.org/10.1007/978-1-4419-7482-2_3

Matthews, S. A., & Yang, T.-C. (2013). Spatial Polygamy and Contextual Exposures (SPACEs): Promoting activity space approaches in research on place and health. *American Behavioral Scientist, 57*(8), 1057–1081. https://doi.org/10.1177/0002764213487345

Milton, S., Pliakas, T., Hawkesworth, S., Nanchahal, K., Grundy, C., Amuzu, A., Casas, J.-P., & Lock, K. (2015). A qualitative geographical information systems approach to explore how older people over 70 years interact with and define their neighbourhood environment. *Health & Place, 36*, 127–133. https://doi.org/10.1016/j.healthplace.2015.10.002

Moore, K., Diez Roux, A. V., Auchincloss, A., Evenson, K. R., Kaufman, J., Mujahid, M., & Williams, K. (2013). Home and work neighbourhood environments in relation to body mass index: The Multi-Ethnic Study of Atherosclerosis (MESA). *Journal of Epidemiology and Community Health, 67*(10), 846–853. https://doi.org/10.1136/jech-2013-202682

Mujahid, M. S., Diez Roux, A. V., Morenoff, J. D., & Raghunathan, T. (2007). Assessing the measurement properties of neighborhood scales: From psychometrics to ecometrics. *American Journal of Epidemiology, 165*(8), 858–867. https://doi.org/10.1093/aje/kwm040

Neely, B., & Samura, M. (2011). Social geographies of race: Connecting race and space. *Ethnic and Racial Studies, 34*(11), 1933–1952. https://doi.org/10.1080/01419870.2011.559262

Peake, L., & Ray, B. (2001). Racializing the Canadian landscape: Whiteness, uneven geographies and social justice1. *The Canadian Geographer/Le Géographe Canadien, 45*(1), 180–186. https://doi.org/10.1111/j.1541-0064.2001.tb01183.x

Perchoux, C., Chaix, B., Cummins, S., & Kestens, Y. (2013). Conceptualization and measurement of environmental exposure in epidemiology: Accounting for activity space related to daily mobility. *Health & Place, 21*, 86–93. https://doi.org/10.1016/j.healthplace.2013.01.005

Petteway, R., Mujahid, M., Allen, A., & Morello-Frosch, R. (2019). Towards a people's social epidemiology: Envisioning a more inclusive and equitable future for social epi research and practice in the 21st century. *International Journal of Environmental Research and Public Health, 16*(20), 3983. https://doi.org/10.3390/ijerph16203983

Powell, J. A., & Cardwell, K. (2013). *Homeownership, wealth & the production of racialized space.* Joint Center for Housing Studies Harvard University. https://lawcat.berkeley.edu/record/1125958

Pulido, L. (2000). Rethinking environmental racism: White privilege and urban development in Southern California. *Annals of the Association of American Geographers, 90*(1), 12–40. https://doi.org/10.1111/0004-5608.00182

Rainham, D., McDowell, I., Krewski, D., & Sawada, M. (2010). Conceptualizing the healthscape: Contributions of time geography, location technologies and spatial ecology to place and health research. *Social Science & Medicine, 70*(5), 668–676. https://doi.org/10.1016/j.socscimed.2009.10.035

Raskind, I. G., Kegler, M. C., Girard, A. W., Dunlop, A. L., & Kramer, M. R. (2020). An activity space approach to understanding how food access is associated with dietary intake and BMI among urban, low-income African American women. *Health & Place, 66*, 102458. https://doi.org/10.1016/j.healthplace.2020.102458

Riva, M., Apparicio, P., Gauvin, L., & Brodeur, J.-M. (2008). Establishing the soundness of administrative spatial units for operationalising the active living potential of residential environments: An exemplar for designing optimal zones. *International Journal of Health Geographics, 7*(1), 43. https://doi.org/10.1186/1476-072X-7-43

Robinson, A. I., & Oreskovic, N. M. (2013). Comparing self-identified and census-defined neighborhoods among adolescents using GPS and accelerometer. *International Journal of Health Geographics, 12*(1), 57. https://doi.org/10.1186/1476-072X-12-57

Roux, A.-V. D. (2007). Neighborhoods and health: Where are we and were do we go from here? *Revue d'epidemiologie et de Sante Publique, 55*(1), 13–21.

Sadler, R. C., Clark, A. F., Wilk, P., O'Connor, C., & Gilliland, J. A. (2016). Using GPS and activity tracking to reveal the influence of adolescents' food environment exposure on junk food purchasing. *Canadian Journal of Public Health, 107*(1), eS14–eS20. https://doi.org/10.17269/CJPH.107.5346

Setton, E., Marshall, J. D., Brauer, M., Lundquist, K. R., Hystad, P., Keller, P., & Cloutier-Fisher, D. (2011). The impact of daily mobility on exposure to traffic-related air pollution and health effect estimates. *Journal of Exposure Science & Environmental Epidemiology, 21*(1), 42–48. https://doi.org/10.1038/jes.2010.14

Shareck, M., Kestens, Y., & Frohlich, K. L. (2014). Moving beyond the residential neighborhood to explore social inequalities in exposure to area-level disadvantage: Results from the interdisciplinary study on inequalities in smoking. *Social Science & Medicine, 108*, 106–114. https://doi.org/10.1016/j.socscimed.2014.02.044

Sharp, G., Denney, J. T., & Kimbro, R. T. (2015). Multiple contexts of exposure: Activity spaces, residential neighborhoods, and self-rated health. *Social Science & Medicine, 146*, 204–213. https://doi.org/10.1016/j.socscimed.2015.10.040

Spielman, S. E., & Yoo, E. (2009). The spatial dimensions of neighborhood effects. *Social Science & Medicine, 68*(6), 1098–1105. https://doi.org/10.1016/j.socscimed.2008.12.048

Vallée, J., Cadot, E., Roustit, C., Parizot, I., & Chauvin, P. (2011). The role of daily mobility in mental health inequalities: The interactive influence of activity space and neighbourhood of residence on depression. *Social Science & Medicine, 73*(8), 1133–1144. https://doi.org/10.1016/j.socscimed.2011.08.009

Vallée, J., Le Roux, G., Chaix, B., Kestens, Y., & Chauvin, P. (2015). The 'constant size neighbourhood trap' in accessibility and health studies. *Urban Studies, 52*(2), 338–357. https://doi.org/10.1177/0042098014528393

Van Wart, S., Tsai, K., & Parikh, T. (2010). Local ground: A paper-based toolkit for documenting local geospatial knowledge. In *ACM symposium on computing for development (DEV)*.

Villanueva, K., Giles-Corti, B., Bulsara, M., McCormack, G. R., Timperio, A., Middleton, N., Beesley, B., & Trapp, G. (2012). How far do children travel from their homes? Exploring children's activity spaces in their neighborhood. *Health & Place, 18*(2), 263–273. https://doi.org/10.1016/j.healthplace.2011.09.019

Widener, M. J., Minaker, L. M., Reid, J. L., Patterson, Z., Ahmadi, T. K., & Hammond, D. (2018). Activity space-based measures of the food environment and their relationships to food purchasing behaviours for young urban adults in Canada. *Public Health Nutrition, 21*(11), 2103–2116. https://doi.org/10.1017/S1368980018000435

Williams, S. A., & Hipp, J. R. (2019). How great and how good?: Third places, neighbor interaction, and cohesion in the neighborhood context. *Social Science Research, 77*, 68–78. https://doi.org/10.1016/j.ssresearch.2018.10.008

Wong, D. W. S., & Shaw, S.-L. (2011). Measuring segregation: An activity space approach. *Journal of Geographical Systems, 13*(2), 127–145. https://doi.org/10.1007/s10109-010-0112-x

York Cornwell, E., & Cagney, K. A. (2017). Aging in activity space: Results from smartphone-based GPS-tracking of urban seniors. *The Journals of Gerontology: Series B, 72*(5), 864–875. https://doi.org/10.1093/geronb/gbx063

Zenk, S. N., Schulz, A. J., Matthews, S. A., Odoms-Young, A., Wilbur, J., Wegrzyn, L., Gibbs, K., Braunschweig, C., & Stokes, C. (2011). Activity space environment and dietary and physical activity behaviors: A pilot study. *Health & Place, 17*(5), 1150–1161. https://doi.org/10.1016/j.healthplace.2011.05.001

Chapter 5
Place, Health, and the Geography of Embodiment: Intergenerational Participatory Research for Representation/ as Resistance in The Ville

(Mis)Representation

Contested Geographies of Place, Health, and Embodiment

> A geographic imperative lies at the heart of every struggle for social justice; if justice is embodied, it is then therefore always spatial, which is to say, part of a process of making a place. (Gilmore, 2002, p. 16)

As Ta-Nehisi Coates (2015) reminds us in regards to racial oppression, "the sociology, the history, the economics, the graphs, the charts, the regressions all land, with great violence, upon the body" (p. 10). Unfortunately, this understanding seems largely absent within much work examining relationships between place and health. Of course, understanding place (e.g., "neighborhoods") and health has become a core focus of public health research and practice (Arcaya et al., 2016; BHPN, n.d.; Diez Roux & Mair, 2010; FRBSF, n.d.; NCHE, n.d.; TCE, n.d.). And there's even a growing interest concerning the physiological embodiment of place-based exposures (Petteway et al., 2019a, b), i.e., how our daily experiences with and within various spatial settings become "biologically embedded" to affect our health and well-being. Yet, much of this work has failed to critically engage who has narrative control in defining "place," whose bodies get talked about, and who gets to do the talking. Research concerned with how place becomes a physiologically embodied reality has relied almost entirely on quantitative survey-based and biometric work that produces, ironically, disembodied accounts of place-embodiment—with credentialed researchers telling statistical stories about people's bodies without affording people an opportunity to speak about their bodies on their own behalf (Petteway et al., 2019a, b). And critically, this work more often than not fails to identify and spatially locate specific place-based exposures, relying upon the same standard crude measures of "place" that plague the place-health field in

© Springer Nature Switzerland AG 2022
R. J. Petteway, *Representation, Re-Presentation, and Resistance*, Global Perspectives on Health Geography, https://doi.org/10.1007/978-3-031-06141-7_5

general—estimating place-based exposures based on ZIP codes or census tracts (see Chaps. 3 and 4; Arcaya et al., 2016; Chaix et al., 2009).

These same concerns and limitations, of course, extend into local health department practice. Standard local health department (LHD) epidemiological practice, e.g., using data based on census tracts or ZIP codes, fails to account for how community residents actually encounter/interact with their daily place-based health exposures/opportunities. People move to and through multiple spatial locations on a daily basis, crossing into and over numerous administrative bounds. Such bounds—and the data produced therein—are thus largely "imaginary" (Petteway, 2017, 2018), as they do not accurately represent or actually define the spatial reality of peoples' place-based health experiences. Unfortunately, most LHD work to date has relied almost exclusively on these sorts of administrative data generated by outside technical experts (see, e.g., CDC, 2020; NAPHSIS, 2020; RWJF, 2020)—with no concerted efforts to engage residents in the data production process. Thus, place-based health data remain very much static, generic, and non-inclusive (e.g., no community participation), which raises concerns about the relevance, reach, and local actionability of these data, as well as concerns around procedural justice and epistemic equity within representations of place and the production of local place-health narratives thereof. Pervading structural and procedural norms within traditional LHD practice preclude expressions of/actively mask community residents' agency, discounting/devaluing their lived place knowledge/expertise, while simultaneously erasing/replacing their place-health experiences with generic and outdated aggregate data (e.g., morbidity, morality, and census data that take years to process for public reporting). These data—and corresponding maps—then come to "represent" place and health for their community, setting the groundwork for related policy and intervention discussions. In this regard, relationships between LHDs (epidemiologists, particularly) and community residents can represent not only the re-inscription of social hierarchy, but the reification of epidemiological surveillance as epistemic violence/erasure.

Moreover, and arguably underlying these concerns, many LHDs face fiscal, political, and jurisdictional limitations that compromise efforts to adequately assess and equitably respond to local place-health concerns. This is particularly true of smaller LHDs that oftentimes do not have (social) epidemiologists and GIS specialists on staff—i.e., they are based at regional or state-levels, not locally—nor the capacity for sustained and authentic community engagement/collaboration for local place-health knowledge co-production. As such, community resident perspective is seldom included as standard in LHD community assessment plans/practice, and youth tend to be excluded entirely. The result is the continuous (re)production of data narratives that materially and symbolically erase community residents' lived geographies of place—replacing them with generic, empiricist, and reductionist spatial (mis)representations detached from history, power, and accountability.

This chapter explores these matters via participatory research with public housing residents in the city of Steubenville, OH. Drawing from the *People's Social Epi Project*, or PSEP (see Chaps. 3 and 4), this chapter introduces the notion of "geographies of embodiment" and details a new method—*X-Ray Mapping*—as a participatory process to more thoroughly examine the placemaking mechanisms that shape residents' placescapes and their embodied place-health geographies therein. In doing so, this chapter highlights the potential value of information and communication technologies (ICTs) in democratizing LHD place-health assessment processes and presents "geographies of embodiment" as a conceptual and analytical frame to improve understanding of how place-based experiences affect health. First, I provide background regarding health and place (mis)representation in the project city to contextualize the project in relation to existing narratives of place. I then detail the X-Ray Mapping process, provide a summary of core findings, and discuss considerations regarding participatory place-embodiment research and the value of ICTs in facilitating resident voice as resistance/re-representation, as well as implications for LHD place-health assessment practice.

Re-Presentation

Participatory Geographies of Embodiment in The Ville

Background

The work described in this Chapter, like that described in Chaps. 3 and 4, is based on the *People's Social Epi Project* (PSEP), which took place Steubenville, OH. The county seat of Jefferson County, and about a 25–30 min drive from Pittsburgh, PA, Steubenville represents somewhat of a quintessential small mid-Western rustbelt city, with the signature post-industrial associated job loss, economic downturn, disinvestment, and infrastructural deterioration. Based on 2010 Census data, of the approximately 70,000 residents of Jefferson County, about 3900 (5.5%) are Black or African American (not counting those who are multiracial). Of that 3900, about 2990 (77%) reside in Steubenville. Similarly, about 60% of Jefferson County's Latino population resides in Steubenville. Moreover, roughly 40% of children under the age of 18 in Steubenville live in poverty, compared to 22% for Jefferson County as a whole. Stated mildly, the city and county populations are not comparable (Petteway, 2016).

As backdrop, the city spends just 3% of its approximately $18.5 million annual budget on "public health and welfare," compared to, for example, 48% on "security of persons and property" (City of Steubenville, 2017). And we know what that means for a City whose police department, in 1997, became the second in the country (following only Pittsburgh) to go under federal consent decree from the Department of Justice Civil Rights Division for brutality, constitutional violations, racially discriminatory practices, false arrests, retaliatory conduct, and corruption (CRLC, n.d.). At the time of this project, the "best" health data available were between 2 and 4 years old, comically rudimentary, arguably racist, and mostly aggregated at the county level (Figs. 5.1 and 5.2)—most of which was not accessible via the LHD website in any format. The data in Fig. 5.1 are from the City's 2012

2012 DEATH STATISTICS

5 year trend of death statistics and demographics:

	2008	2009	2010	2011	2012
White Male	356	358	320	339	374
Non-White Male	20	31	24	27	24
White Female	369	394	353	375	323
Non-White Female	30	24	19	29	25
TOTAL DEATHS	**775**	**807**	**716**	**770**	**746**
Steubenville Residents	255	277	246	300	238
Jeff. County (Including Steub.)	544	584	516	575	501
Out of Jefferson County-in Ohio	82	78	59	59	78
Out of State	149	145	141	136	167

Fig. 5.1 Summary death data from the 2012 Steubenville Health Department Annual Report. Excerpt from the 2012 Steubenville Health Department Annual Report showing raw counts of death data. Note the use of "White" and "Non-White" as racial categories

Health Measure	Steubenville/Jefferson County, 2010 (2012)	Pittsburgh/Allegheny County, 2010 (2012)	U.S., 2010 (2012)
Heart Disease Mortality	275.2	183.6	193.6
Cancer Mortality	188.4	184.8	186.2
Stroke Mortality	42.9	38.9	41.9
Chonic Lower Respiratory Disease Mortality	64.8	36.6	44.7
Unintentional Injury (Accident) Mortality	46.6	43.2	39.1
Fair or Poor Health	(23%)	(13%)	(17%)
Adult Obesity	(37%)	(29%)	(28%)
Adult Smoking	(26%)	(20%)	(21%)
Uninsured	(13%)	(10%)	(18%)

Fig. 5.2 Summary health data for Jefferson County, OH, where Steubenville is located. Table summarizing various health indicators as contained in the 2010 and 2012 Robert Wood Johnson Foundation's County Health Rankings

Annual Report (SHD, 2012). The report did not include a single standardized health indicator, only raw counts. It also did not include any indicators on community environmental conditions or any neighborhood-specific data. And it allocated more space to plumbing (I'm not making this up) than to maternal and child health.

Nobody within the city government—including the health commissioner, the city manager, and the mayor—knew what the average life expectancy was for city residents overall, let alone for low-income and communities of color residing in public housing (note: it was 69 years for the census tract where PSEP participants lived). Moreover, there had never been—and still hasn't been—any comprehensive and systematic assessment of place-based health exposures, risks, and opportunities. For example, there was and remains precisely zero data regarding place-based measures of social determinants of health for the city, e.g., food environment, tobacco and alcohol retail environment, greenspace and tree canopy, sidewalk quality and pedestrian safety, transportation, housing conditions, and so on. And as a city that served a pivotal role in the development of the US Environmental Protections Agency's guidelines regarding air pollution (Dockery et al., 1985; Laden et al., 2006; Lepeule et al., 2012)—and still remains in the bottom 10 metro areas nationwide (ALA, 2021), with bottom 10 percentile quality near schools (PERI, 2021)—there is zero environmental health data available via the LHD.

Given the data that was available at the time, one could only anticipate that if Jefferson County was faring worse than Pittsburgh/Allegheny County (even given its horrendous health record for low-income and communities of color, see, e.g., Howell et al., 2019), and Steubenville is where most people of color in Jefferson County reside—and it has twice the poverty rate—then things might need some immediate attention. It's also worth noting here that during the PSEP, the City and County were in the process of merging health departments, with Steubenville's health department being dissolved and the county essentially taking over. The largest city in the county—with double the poverty rates and an overwhelming majority of the county's Black and Latino populations—now has no direct LHD representation. Residents are left without any clear information regarding the health of their communities.

Also unfolding during PSEP was the City's comprehensive planning process. Amongst many notable aspects of the plan regarding community inclusion and (mis)representation, Fig. 5.3 shows the manner in which PSEP participants' entire community was viewed by those in city planning—as lacking "a sense of place." In developing the PSEP, it was clear that the residents of PSEP participants' broader housing community were not included in the assessment and planning process—that the City declared that their community lacked a sense of place without having engaged residents in a conversation to hear their perspectives regarding their "place." As such, and as noted in the discussion of Chap. 3, residents viewed PSEP as a way to counter this misrepresentation and resist what they viewed as the City's effort to "wipe [them] off the map." We accordingly sought ways to re-map and re-present "place" from residents' lived and embodied perspectives, and reached out to the

Fig. 5.3 Excerpt from 2013 Steubenville Comprehensive Plan. Section of Steubenville's 2013 Comprehensive Plan describing "conditions" and "opportunities" present within PSEP participants' broader neighborhood. The highlighted section reads: "lacks sense of place"

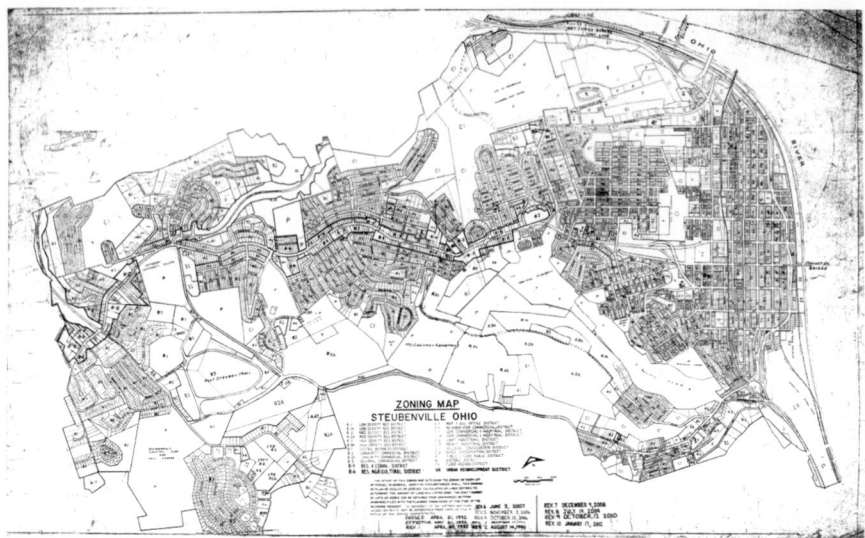

Fig. 5.4 Zoning map of Steubenville, OH, 2012. Zoning map of Steubenville, OH that was provided by the City planning department. It is the only map they were willing to share for the PSEP

City planning department to gain access to some GIS maps and shapefiles. Figure 5.4 is either all they had or all they were willing to share. Either way, the message was clear. And the health department had no maps or shapefiles to share at all.

X-Ray Mapping Process

X-Ray Mapping built on two previous participatory methods used during PSEP: photovoice (see Chap. 3 and Petteway, 2019 for details), and activity space mapping (see Chap. 4). For a more detailed background on the conceptual roots of X-Ray Mapping, and the conceptual roots and empirical groundings of embodiment in place-health research, see Ruglis (2011) and Petteway and colleagues (2019a, b). Here, participants used 8.5" by 11" worksheets with a basic body outline with dorsal and ventral representation (i.e., front and back) to identify body areas that they believed were affected by each of their places (Figs. 5.5 and 5.6). For each photo-place identified via photovoice and activity space mapping, they created an X-Ray Map using stickers to indicate the areas of their body they perceived were affected by that particular place. That is, each photo-place had a corresponding X-Ray Map to represent participants' perceptions of place-embodiment. This was done over the course of two meetings. Participants expressed a desire to continue the color-coding scheme from the activity space mapping. Here, green represented healthy/good/positive body effects, red represented unhealthy/bad/negative body effects, and yellow represented both. Participants were free to use as many stickers as they believed necessary to capture all of their perceived place-embodiment effects for each place, such that each X-Ray Map could contain multiple positive and negative effects (e.g., positive heart, negative brain, and negative back) and each body area could have

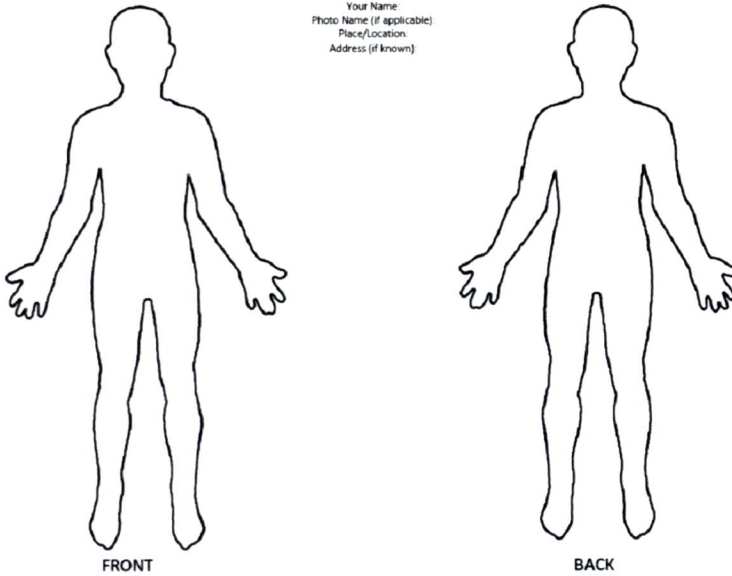

Fig. 5.5 X-Ray Mapping worksheet. X-Ray Mapping worksheet used by participants to indicate how and where their daily places affected their bodies

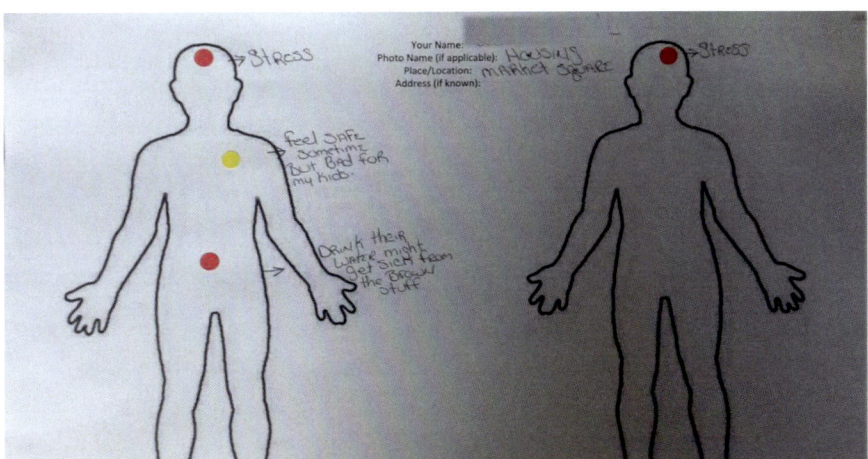

Fig. 5.6 Example of completed X-Ray Map. Completed X-Ray Map from an adult participant, describing the different ways—and places—that their housing unit affects their body

multiple stickers of the same or different colors (e.g., two positive and three nega-tive brain effects). Participants were instructed to use the back of their X-Ray Map worksheets to write a brief description/narrative explaining their place-embodiment representations.

Each X-Ray Map was reviewed to complete simple counts and frequencies of: (1) place-embodiment geographic locations based on the five overall PSEP place-domains of *Home*, *Neighborhood*, *School/Work*, *Leisure/Social*, and *Transition* (see

Chaps. 3 and 4), (2) place-embodiment physiologic locations (e.g., heart, brain, stomach), and (3) type of perceived place-embodiment effect (i.e., positive, negative, both). This was done for each individual participant separately. Once individual place-embodiment tabulations were completed, results were aggregated for youth and parents separately. Aggregate summary tables were produced for overall adult and youth place-embodiment data, as well as domain-specific adult and youth place-embodiment data. Qualitative comparisons were made between aggregate youth and aggregate adult X-Ray data. Summary infographics were developed to visually represent place-embodiment among adult and youth participants. All X-Ray Map data was then mapped on the *Local Ground* web-based community mapping platform (Van Wart et al., 2010), enabling geographic visualization and qualitative comparison of adult and youth "geographies of embodiment."

X-Ray Mapping Findings

Youth completed a total of 45 X-Ray Maps, while adults completed 23. Overall, youth indicated that their daily places positively and/or negatively affected 20 different body areas across the 5 place-domains, with a total of 107 perceived body effects across the 20 body areas (Fig. 5.7). Adults identified 12 body areas, with a total of 87 perceived body effects (Fig. 5.8).

X-Ray Mapping results revealed that 49% of youth place-embodiment locations were spatially outside of their residential census tract—with 75% of *positive* place-embodiment locations outside, and 66% of *negative* place-embodiment locations inside (Fig. 5.9). Among adults, positive and negative place-embodiment locations were about evenly distributed inside and outside of their residential census tract

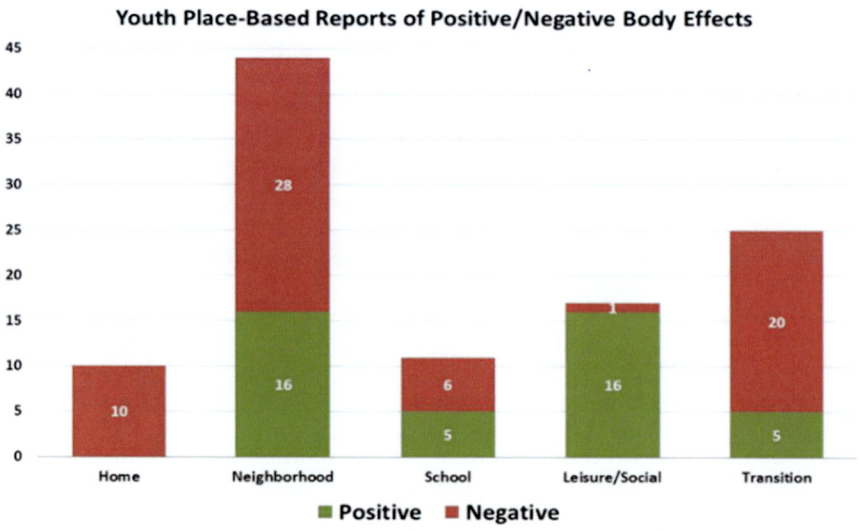

Fig. 5.7 Youth place-embodiment effects by place-domain. Chart summarizing youths' reported place-embodiment effects based on the PSEP's 5 broad place-domains.

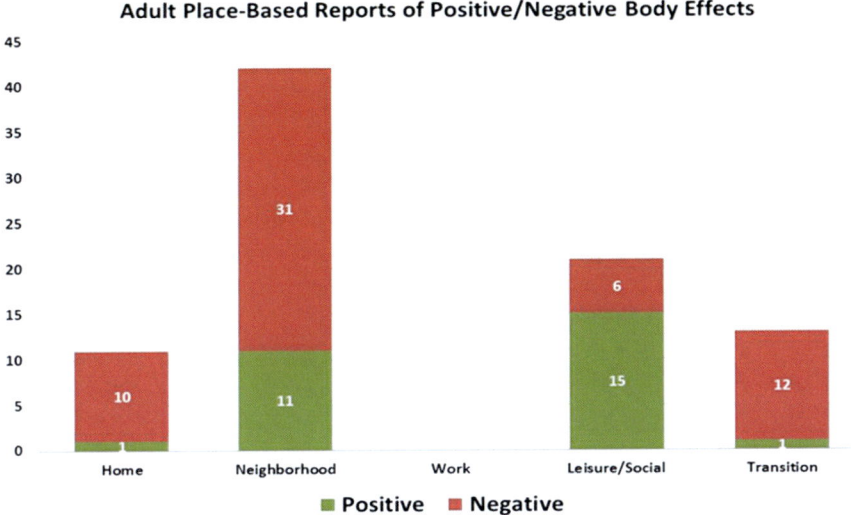

Fig. 5.8 Adult place-embodiment effects by place-domain. Chart summarizing youths' reported place-embodiment effects based on the PSEP's 5 broad place-domains.

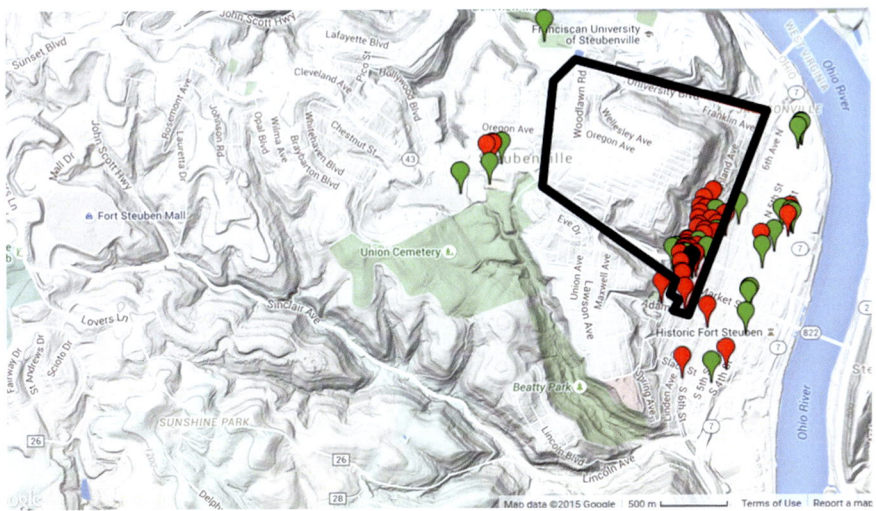

Fig. 5.9 Youth positive/negative place-embodiment locations. Youth reported perceptions of place-embodiment for specific locations they encounter. Green markers represent locations youth perceived as having a positive body-effect, while represent red markers represent a perceived negative body-effect. The black polygon is an outline of the census tract in which the participants' housing community is located. The black marker is their housing location

(Fig. 5.10). Overall, 67% of youth and adult positive place-embodiment locations were outside of their residential census tract.

Figures 5.11 and 5.12 graphically represent adult and youth "geographies of embodiment." The five place-domains are represented by the color-coded symbols

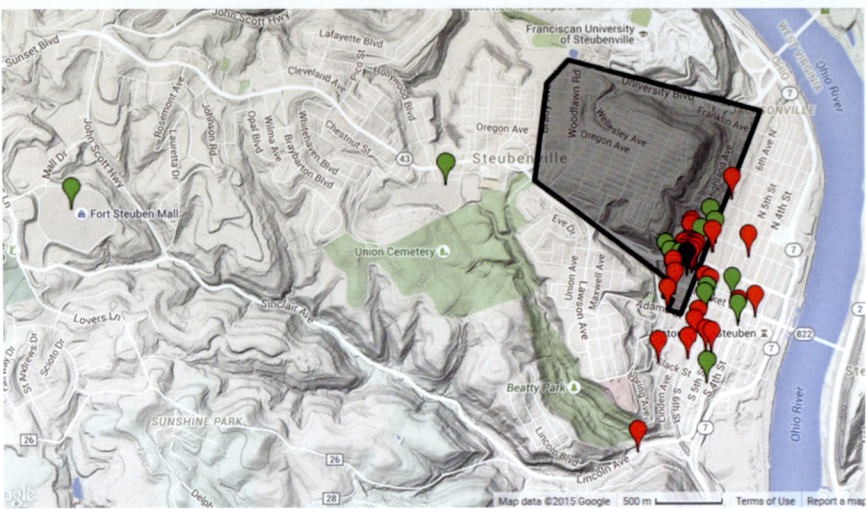

Fig. 5.10 Adult positive/negative place-embodiment locations. Adult reported perceptions of place-embodiment for specific locations they encounter. Green markers represent locations youth perceived as having a positive body-effect, while represent red markers represent a perceived negative body-effect. The black polygon is an outline of the census tract in which the participants' housing community is located. The black marker is their housing location

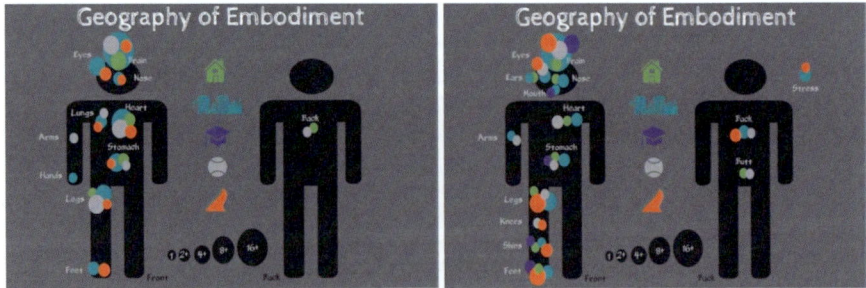

Fig. 5.11 Adult and youth "Geographies of Embodiment" by place domain. *Geography of Embodiment for Adults (L) and Youth (R)*. Green represents the "Home" place-domain; Blue represents "Neighborhood"; Purple represents "School/Work"; Grey represents "Leisure/Social"; and Orange represents "Transition". The size of the circle reflects the number of times a specific body area was identified as being affected. This figure includes both positive and negative place-embodiment perceptions

and embodiment circles. The size of the circles corresponds to the number of times a specific body area was identified as being affected, either positively or negatively, within that particular place-domain. Green represents the "Home" place-domain; blue represents the "Neighborhood" place-domain; purple represents the "School/ Work" place-domain; grey represents the "Leisure/Social" place-domain; and orange represents the "Transition" place-domain. Figure 5.12 shows adult and youth

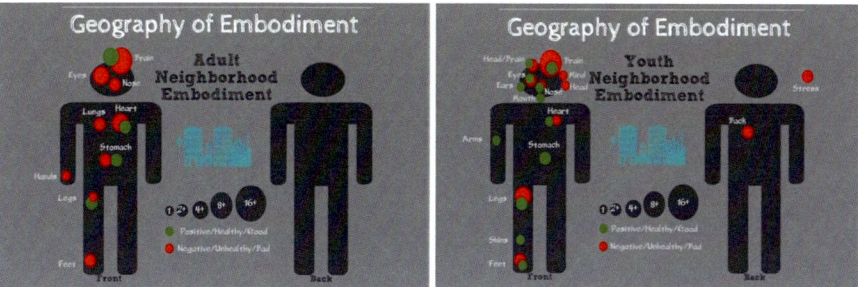

Fig. 5.12 Adult and youth positive/negative embodiment for the "neighborhood" place domain. *Geography of Embodiment for Adults (L) and Youth (R)*. Green represents perceived positive/health/good place-embodiment effects; Red represents perceived negative/unhealthy/bad place-embodiment effects. The size of the circle reflects the number of times a specific body area was identified as being affected

place-embodiment perceptions specifically for the neighborhood place-domain, with representation of positive and negative body-effects. Here, indicated positive effects are represented in green, and negative effects are represented in red.

Resistance

Discussion: Participatory Place-Health Geographies as Embodied Resistance

The goal of this chapter was to introduce a new method for examining notions of embodiment within place-health, highlighting the value of participatory processes that center community voice and experience. In doing so, I also sought to illustrate a potential process through which LHDs can more thoroughly include resident perspective and knowledge within standard community health assessment practices, especially as related to considerations of "place"—the intention being to democratize standard practices which commonly serve as mechanisms of epistemic erasure, place misrepresentation, and spatial stigma. On these matters, the work presented here suggests five broad considerations and implications for place-health research and LHD practice (see Petteway et al., 2019a, b for a more detailed discussion of specific data findings).

First, as discussed in Chap. 4 regarding related data, findings illustrate the importance of moving away from administratively defined notions of "place" when examining health, and embracing relational and activity space approaches (Cagney et al., 2020; Chaix et al., 2009; Cummins, 2007; Cummins et al., 2007; Kwan, 2009; Perchoux et al., 2013). Figures 5.9 and 5.10 make it clear that standard administrative definitions of participants' place-health exposures/encounters would have not only led to misspecification, but would have entirely misrepresented their actual place-health experiences and geographies. As I have written elsewhere (Petteway, 2018), residents' lives are not bounded or spatially defined by "imaginary lines" like census tracts and ZIP codes. Most place-embodiment research to date, however, has ignored mobility, focusing only on place-health exposures/risks as defined and bounded by peoples' residential census tracts (Petteway et al., 2019a, b), oftentimes assuming uniformity of exposure across the entire geographic area. As notable in looking at Figs. 5.9 and 5.10, such approaches should not only raise deep concerns regarding misestimation of place effects, but more fundamentally, misrepresentations of place-embodiment.

Second, of particular significance here, is the spatially specific, qualitative and subjective assessment of place-embodiment. Research has shown that both objective and subjective measures of place matter for health (Lin & Moudon, 2010; Pruitt et al., 2012; Schulz et al., 2013; Shareck & Ellaway, 2011; Tilt et al., 2007; Weden et al., 2008; Wen et al., 2006), which should raise questions regarding why this PSEP work might represent the first qualitative and spatially specific—and intergenerational—work to examine place-embodiment in the place-health literature. For PSEP, spatially locating perceptions of embodiment as related to specific physical and social environmental factors allowed participants to tell a story of place-embodiment within which potential *pathways of embodiment* can more readily be discerned (Krieger, 2001). These pathways—the processes and mechanisms through which current and historic societal arrangements of power, privilege, and opportunity shape and organize contexts and conditions of embodiment—are seldom (if ever) explicated within place-health research. These pathways are very much a part/expression of placemaking mechanisms that, as discussed in detail in Chap. 3 (and Chap. 6), must be accounted for to better contextualize and respond to place-based inequities.

Given the spatial and physiologic range of embodiment effects reported (Figs. 5.9, 5.10, 5.11 and 5.12), there is implication of a broad range of local/regional policies, processes, and practices that shape daily living conditions, experiences, and exposures that become physiologically embodied by residents. This is in stark contrast to much place-embodiment work to date that has both failed to identify and spatially locate specific place attributes implicated in place-embodiment, and foregone attempts to uncover elements of the local context (i.e., pertinent to the samples' place-embodiment prospects) that are on or might constitute potential pathways of embodiment, i.e., ignoring the placemaking mechanisms that (re)produce the social and material contexts of place that we embody. By identifying specific attributes/aspects of place contexts and their relations to embodiment, and by elucidating a more robust spectrum of potential social and political processes and practices that shape place-embodiment patterns, we improve our ability to answer fundamental questions in regards to: (1) who and what shapes/determines distributions and patterns of underlying health opportunities, exposures, and risks in these locally experienced spaces of embodiment; (2) what these distributions and patterns can tell us about agency and accountability at the local and regional level; and (3) what is at stake in, and what is the value of, telling spatially and jurisdictionally specific stories of embodiment. The approach taken to the work discussed here allows for these questions to be meaningfully engaged within participants' daily place-health contexts.

Third, this project highlights the potential for, and value of, participatory LHD place-health assessment practices that are community-centered and community-led, which very much aligns with the much-touted principles of health equity that LHDs espouse, and the revised "10 Essential Public Health Services" (PHNCI, 2020). The notion of health equity circulates heavily within public health research and practice. Yet, missing from the conversation are considerations for epistemic equity, and procedural, data, and distributive justice. What does it mean to "advance" and "center" equity within LHD practice if/when standard community assessment practices exclude the voices of those most burdened by health inequities? And if, in most contexts, data regarding place and health are based on secondary analyses of administrative morbidity, mortality, and census data—done in isolation by analysts in air-conditioned offices/cubicles—how is that remotely an equitable—or accurate, even—representation of residents' daily place-health contexts? Such data are often years old by the time they are released for analyses, and most LHDs don't even make such data and subsequent analyses publicly accessible/downloadable. At best, clicking around on most LHD websites might lead you to a static PDF document from 2 years ago, and there's nothing to guarantee data will be community-specific, e.g., "neighborhood."

For this project, as noted above, there was literally no data available regarding residents' place-health contexts—neither specifically for their housing community and surrounding neighborhood, nor their city as a whole. And the data that was available rendered their lived realities of place invisible or otherwise misrepresented their place—excluding them from the larger discourse and narrative production related to "place," health, and opportunity in their community. Existing data was rudimentary at best, and arguably racist. And City administrators quite literally mapped/described their entire community as lacking a "sense of place"—in essence carrying out, in the words of McKittrick (2011, p. 954), "a willful destruction of a Black sense of place." Such destructions—as symbolic/epistemic erasures and violence—are naturalized because, as articulated by McKittrick (2011, p. 954), "blackness is recognizably placeless and degraded and therefore justifiably without, which is… the commonsense outcome of our analytical queries." The work described here, as noted in Chap. 3, was thus envisioned as an avenue to (re)shape local discourse, a method to represent/re-present important community spaces, and a tactic to resist/counter efforts to, in the words of residents, "wipe us off the map." And in many ways, it represented the first time their place-health contexts were included "on the map" at all. This work accordingly suggests a potential model for how to engage the important considerations Alang in colleagues (2021) raise in regards to White supremacy within/in relation to the 10 Public Health Essential Services—and especially in regards to items #1, #2, and #4, related to community assessment and policy development (PHNCI, 2020).

The participatory X-Ray Mapping process allowed residents to identify specific spatial locations with specific exposures, risks, and opportunities that implicate specific policy domains and social action targets. And they were able to generate representations/re-presentations of place and health—and place-embodiment—that reflected their lived experiences of place. In this way, the process allowed for the production of new place-health knowledge that not only is much more nuanced and spatially specific, but also socially relevant and actionable. As such, this work suggests that the use of participatory tools like X-Ray Mapping can help center resident voice within LHD community assessment processes and offer at least a partial response to concerns around procedural and epistemic justice in the production of place-health narratives. Moreover, through mapping "geographies of embodiment" via tools like X-Ray Mapping, we can gain greater insight into *how* place affects health—adding depth and texture to place-embodiment research that to date has been almost exclusively quantitative and based on surveys and biometrics. These gains can greatly enhance LHD efforts to assess/address elements of local social, political, economic, and environmental contexts that constitute expressions of social and spatial inequality—perhaps offering a way to bridge administrative and community-generated data to better represent and address factors that fundamentally drive or constrain residents' place-based health opportunities/risks.

Fourth, this work was greatly facilitated by the technological and procedural affordances of ICTs, namely, smartphones and the *Local Ground* web-based mapping platform. The use of ICTs for/within health and human development work is informed by a rich body of conceptual and empirical work, commonly referred to as ICT for development, or ICTD. Core to the decision to use ICTs in this project were notions of "liberation technology," "deliberation technology," and "small data." Liberation technology has been defined as any form of ICT that "can expand political, social, and economic freedom" (Diamond, 2010, p. 60). This includes, for example, ICT use to increase government transparency and accountability, to organize and mobilize for social action, to generate and disseminate independent news (e.g., "citizen journalists"), and to simplify and deepen civic participation. As described by Diamond (2010), "liberation technology enables citizens to report news, expose wrongdoing, express opinions, mobilize protest, monitor elections, scrutinize government, deepen participation, and expand the horizons of freedom" (p. 70). It is mostly concerned with the role of ICTs in organizing and amplifying the voice of dissident and/or marginalized groups *during* political struggles/transformations. The related notion of "deliberation technology" was introduced to reframe liberation technology and expand focus to examine the role of ICTs *after* political struggles/transformation as well (Pfister & Godana, 2012). Accordingly, *de*liberation technologies (p. 2):

> …facilitate not just information circulation, but discussion and debate. Deliberation technologies focus not just on the hardware of communication, but on the software and the practices that support a broad-based conversation amongst affected citizens. Deliberation technologies do not serve specific and episodic goals, but focus on cultivating sites of sustained communication.

As described by D'Ignazio et al. (2014, p. 116), *Small Data* is, a practice owned and directed by those who are contributing the data... The essence of Small Data is that such communities may not just participate in, but can actually initiate and drive such data investigations towards the better understanding of an important local issue.

They suggest, specifically in regard to investigating environmental factors, that "a bottom-up, participatory, grassroots approach to... data collection addresses the key issues of inclusion, accountability, and credibility, by building public participation into the data lifecycle" (p. 116). For the work presented here, the combination of these three concepts—liberation technology, deliberation technology, and small data—suggested an opportunity to harness ICTs for engaging residents in a deeply participatory, deliberative, and potentially actionable inquiry of their lived and embodied place-health geographies. If social and political action is ultimately what is necessary to alter/reconfigure placescapes and address place-based health inequities, it seems critical that researchers/practitioners consider the technological, procedural, and translational affordances of participatory ICTs to amplify potential impacts of local place-health work—such that those experiencing the embodied consequences of inequitable place-health contexts have the opportunity to inform, lead, narrate, and act upon the work.

Fifth, and in sum, participants' work illustrates the value of anchoring place-health assessment processes in residents' lived experience, and more deeply/expressly engaging critical theory, feminist, and Black feminist orientations to place-health geography knowledge production. Part of what animated the work presented here were concerns regarding lack of representation and the misrepresentation of residents' place-health exposures and experiences within local health discourse and practice. The choice to use exclusively participatory methods, including X-Ray Mapping, was a deliberate one to "center the margins" (hooks, 2000), honor residents' "voice" (Ford & Airhihenbuwa, 2010), generate counterstories (Delgado, 1989; Solórzano & Yosso, 2002), and facilitate what Freire refers to as "critical consciousness" (Freire, 2000). Participants were accordingly able to draw from their situated—socially and spatially—knowledges to offer a (counter)narrative of their place-health geographies—a critical rearticulation of their community health contexts as (mis)represented via dominant discourse(s) circulating in their city, and/as circulated *by* their *City*. In this way, participants' work reflects what Foucault (1980) called "productive power"—they *produced* their own place-health geographies, thereby producing *their* place into existence as counternarrative to that produced by traditional place-health research and city administrators in an acts of "repressive power." As noted above, if LHD practice is to be guided by the 10 Public Health Essential Services, considerations of White supremacy—as evident within dominant public health discourses/practice (Alang et al., 2021; Petteway, 2021)—must be expounded and engaged to truly advance equity.

In this capacity, participants' place-health representations via these unique embodiment maps not only present as new data but as counterstories—as counternarrative to existing tropes and (mis)representations of their community and experiences therein. As discussed elsewhere (de Leeuw & Hawkins, 2017; Gislason et al., 2018; Peterle, 2019; Swords et al., 2019; Velasco et al., 2020; Bates et al., 2018), these sorts of creative and participatory processes not only serve to reimagine and remake place by generating new spatial knowledges, but also by articulating new spatial narratives and representations of place. And as noted by Neely and Samura (2011), "relations of power intersect with and may force the movement or placement of people; they also inform the knowledge produced within and about particular spaces" (p. 1936). In the context of this project, participants, as mostly Black residents of a racialized low-income housing community, created "geographies of embodiment" that constitute expressions of power within the production of space and knowledge about place and health—expressions that can inform LHD practice and help reshape local narratives of their "place."

The X-Ray Mapping methodology, and other participatory intergenerational approaches that draw upon peoples' knowledge of local contexts and their lived and embodied experience of "place," could represent and encourage novel approaches to community assessment for LHD strategizing and comprehensive city planning. Also, taking advantage of the increasing availability and utility of ICTs could further enhance the value and extend the reach of methodologies like X-Ray Mapping. The research presented here made use of a web-based multimedia-enabled community mapping platform, thus enabling participants' geographies of embodiment to be digitally mapped and readily shared and distributed. Such ICTs, appropriately designed and deployed (Avgerou, 2010; Burrell & Toyama, 2009; Dearden, 2012), raise the prospect of population-wide assessment of place, embodiment, and health relationships in both research and practice, which could be of particular value for under-resourced and under-staffed LHDs that have little-to-no history of community engagement and/or inclusive data practices. Incorporating ICTs to acknowledge and facilitate residents' agency and collective power in order to pursue collaborative place-health data practices would do procedural, epistemic, and analytic justice to LHD assessment and mapping practices, and would anchor the data in the lived realities of those actually embodying the places on the maps. This can render a more humanized and actionable data narrative, while simultaneously structuring meaningful opportunities for civic engagement and expanding mechanisms for public accountability. This could prove pivotal to local assessment and action efforts that frequently hinge upon local politics and agenda setting (Castrucci et al., 2015; Mowat & Chambers, 2012; Smylie et al., 2012).

Conclusion

Through mapping "geographies of embodiment" via participatory methods like *X-Ray Mapping*, we can gain greater insight into what is embodied (i.e., specific experiences/exposures), when (i.e., temporally specific), and where (i.e., spatially specific). These gains can improve development of quantitative place-health metrics and greatly enhance efforts to uncover/intervene on the "pathways of embodiment"—specifically, those elements of local social, political, economic, and environmental contexts that constitute expressions of social inequality. Anchored in

X-Ray Mapping, and facilitated by the use of ICTs, the geographies of embodiment concept is responsive to existing limitations within place-embodiment research, offering a way to reframe and re-approach our work. It is not only capable of revealing general patterns of place-embodiment within a particular community but can reveal specific place attributes within those patterns that directly or indirectly implicate local policies and practices that shape daily social and physical environments. And importantly, it could facilitate movement toward a more collaborative, inclusive, and procedurally and epistemically justice data practice within LHD efforts to identify and respond to community place-health concerns, related to embodiment or otherwise.

References

ALA. (2021). *Most polluted cities*. American Lung Association, State of the Air. /research/sota/city-rankings/most-polluted-cities.

Alang, S., Hardeman, R., Karbeah, J., Akosionu, O., McGuire, C., Abdi, H., & McAlpine, D. (2021). White supremacy and the core functions of public health. *American Journal of Public Health, 111*(5), 815–819. https://doi.org/10.2105/AJPH.2020.306137

Arcaya, M. C., Tucker-Seeley, R. D., Kim, R., Schnake-Mahl, A., So, M., & Subramanian, S. V. (2016). Research on neighborhood effects on health in the United States: A systematic review of study characteristics. *Social Science & Medicine, 168*, 16–29. https://doi.org/10.1016/j.socscimed.2016.08.047

Avgerou, C. (2010). Discourses on ICT and development. *Information Technologies and International Development, 6*(3), 1–18.

Bates, L. K., Towne, S. A., Jordan, C. P., Lelliott, K. L., Bates, L. K., Towne, S. A., Jordan, C. P., Lelliott, K. L., Johnson, M. S., Wilson, B., Winkler, T., Brand, A. L., Corbin, C. N. E., Miller, M. J., Koh, A., Freitas, K., & Roberts, A. R. (2018). Race and spatial imaginary: Planning otherwise/Introduction: What shakes loose when we imagine otherwise/She made the vision true: A journey toward recognition and belonging/Isha Black or Isha White? Racial identity and spatial development in Warren County, NC/Colonial City design lives here: Questioning planning education's dominant imaginaries/Say its name – Planning is the white spatial imaginary, or reading McKittrick and Woods as planning text/Wakanda! Take the wheel! Visions of a black green city/If I built the world, imagine that: Reflecting on world building practices in black Los Angeles/Is Honolulu a Hawaiian Place? Decolonizing cities and the redefinition of spatial legitimacy/Interpretations & imaginaries: Toward an instrumental black planning history. *Planning Theory & Practice, 19*(2), 254–288. https://doi.org/10.1080/1464935 7.2018.1456816

BHPN. (n.d.). *Building healthy places network*. Build Healthy Places Network. Retrieved June 9, 2021, from https://buildhealthyplaces.org/

Burrell, J., & Toyama, K. (2009). What constitutes good ICTD research? *Information Technologies and International Development, 5*(3), 13.

Cagney, K. A., York Cornwell, E., Goldman, A. W., & Cai, L. (2020). Urban mobility and activity space. *Annual Review of Sociology, 46*(1), 623–648. https://doi.org/10.1146/annurev-soc-121919-054848

Castrucci, B. C., Rhoades, E. K., Leider, J. P., & Hearne, S. (2015). What gets measured gets done: An assessment of local data uses and needs in large urban health departments. *Journal of Public Health Management and Practice, 21*, S38–S48. https://doi.org/10.1097/PHH.0000000000000169

CDC. (2020). *500 cities project: Local data for better health.* https://www.cdc.gov/500cities/index.htm

Chaix, B., Merlo, J., Evans, D., Leal, C., & Havard, S. (2009). Neighbourhoods in eco-epidemiologic research: Delimiting personal exposure areas. A response to Riva, Gauvin, Apparicio and Brodeur. *Social Science & Medicine, 69*(9), 1306–1310. https://doi.org/10.1016/j.socscimed.2009.07.018

City of Steubenville. (2017). *Comprehensive financial report for the year ending December 31, 2016.* City of Steubenville.

Coates, T.-N. (2015). *Between the world and me.* Spiegel & Grau.

CRLC. (n.d.). *U.S. v. City of steubenville.* Civil Rights Litigation Clearinghouse. Retrieved June 8, 2021, from https://www.clearinghouse.net/detail.php?id=1035

Cummins, S. (2007). Commentary: Investigating neighbourhood effects on health – Avoiding the "Local Trap". *International Journal of Epidemiology, 36*(2), 355–357. https://doi.org/10.1093/ije/dym033

Cummins, S., Curtis, S., Diez-Roux, A. V., & Macintyre, S. (2007). Understanding and representing 'place' in health research: A relational approach. *Social Science & Medicine, 65*(9), 1825–1838. https://doi.org/10.1016/j.socscimed.2007.05.036

D'Ignazio, C., Warren, J., & Blair, D. (2014). The role of small data for governance in the 21st century. In *Governança Digital* (pp. 115–129). Universidade Federal Rio Grande do Sul.

de Leeuw, S., & Hawkins, H. (2017). Critical geographies and geography's creative re/turn: Poetics and practices for new disciplinary spaces. *Gender, Place & Culture, 24*(3), 303–324. https://doi.org/10.1080/0966369X.2017.1314947

Dearden, A. (2012). *See no evil?: Ethics in an interventionist ICTD, 46.* https://doi.org/10.1145/2160673.2160680

Delgado, R. (1989). Storytelling for oppositionists and others: A plea for narrative. *Michigan Law Review, 87*(8), 2411–2441. https://doi.org/10.2307/1289308

Diamond, L. (2010). Liberation technology. *Journal of Democracy, 21*(3), 69–83. https://doi.org/10.1353/jod.0.0190

Diez Roux, A. V., & Mair, C. (2010). Neighborhoods and health: Neighborhoods and health. *Annals of the New York Academy of Sciences, 1186*(1), 125–145. https://doi.org/10.1111/j.1749-6632.2009.05333.x

Dockery, D. W., Ware, J. H., Ferris, B. G., Glicksberg, D. S., Fay, M. E., Spiro, A., & Speizer, F. E. (1985). Distribution of forced expiratory volume in one second and forced vital capacity in healthy, white, adult never-smokers in six U.S. cities. *American Review of Respiratory Disease, 131*(4), 511–520. https://doi.org/10.1164/arrd.1985.131.4.511

Ford, C. L., & Airhihenbuwa, C. O. (2010). The public health critical race methodology: Praxis for antiracism research. *Social Science & Medicine, 71*(8), 1390–1398. https://doi.org/10.1016/j.socscimed.2010.07.030

Foucault, M. (1980). *Power/knowledge: Selected interviews and other writings, 1972–1977* (C. Gordon, Ed.). Pantheon Books.

FRBSF. (n.d.). *Healthy communities.* Federal Reserve Bank of San Francisco: Healthy Communities Initiative. Retrieved June 9, 2021, from https://www.frbsf.org/community-development/initiatives/healthy-communities/

Freire, P. (2000). *Pedagogy of the oppressed* (30th Anniv ed.). Continuum.

Gilmore, R. W. (2002). Fatal couplings of power and difference: Notes on racism and geography. *The Professional Geographer, 54*(1), 15–24. https://doi.org/10.1111/0033-0124.00310

Gislason, M. K., Morgan, V. S., Mitchell-Foster, K., & Parkes, M. W. (2018). Voices from the landscape: Storytelling as emergent counter-narratives and collective action from northern BC watersheds. *Health & Place, 54*, 191–199. https://doi.org/10.1016/j.healthplace.2018.08.024

hooks, b. (2000). *Feminist theory: From margin to center.* Pluto Press.

Howell, J., Goodkind, S., Jacobs, L., Branson, D., & Miller, L. (2019). *Pittsburgh's inequality across gender and race* (Gender Analysis White Papers). City of Pittsburgh's Gender Equity Commission.

Krieger, N. (2001). Theories for social epidemiology in the 21st century: An ecosocial perspective. *International Journal of Epidemiology, 30*(4), 668–677. https://doi.org/10.1093/ije/30.4.668

Kwan, M.-P. (2009). From place-based to people-based exposure measures. *Social Science & Medicine, 69*(9), 1311–1313. https://doi.org/10.1016/j.socscimed.2009.07.013

Laden, F., Schwartz, J., Speizer, F. E., & Dockery, D. W. (2006). Reduction in fine particulate air pollution and mortality: Extended follow-up of the Harvard six cities study. *American Journal of Respiratory and Critical Care Medicine, 173*(6), 667–672. https://doi.org/10.1164/rccm.200503-443OC

Lepeule, J., Laden, F., Dockery, D., & Schwartz, J. (2012). Chronic exposure to fine particles and mortality: An extended follow-up of the Harvard six cities study from 1974 to 2009. *Environmental Health Perspectives, 120*(7), 965–970. https://doi.org/10.1289/ehp.1104660

Lin, L., & Moudon, A. V. (2010). Objective versus subjective measures of the built environment, which are most effective in capturing associations with walking? *Health & Place, 16*(2), 339–348. https://doi.org/10.1016/j.healthplace.2009.11.002

McKittrick, K. (2011). On plantations, prisons, and a black sense of place. *Social & Cultural Geography, 12*(8), 947–963. https://doi.org/10.1080/14649365.2011.624280

Mowat, D., & Chambers, C. (2012). Producing more relevant evidence: Applying a social epidemiology research agenda to public health practice. In P. O'Campo & J. R. Dunn (Eds.), *Rethinking social epidemiology* (pp. 305–325). Springer. https://doi.org/10.1007/978-94-007-2138-8_15

NAPHSIS. (2020). *USALEEP: Neighborhood life expectancy project*. Naphsis. https://www.naphsis.org/usaleep

NCHE. (n.d.). *National collaborative for health equity: Place matters initiative*. Retrieved September 30, 2015, from http://nationalcollaborative.org/?q=node/37

Neely, B., & Samura, M. (2011). Social geographies of race: Connecting race and space. *Ethnic and Racial Studies, 34*(11), 1933–1952. https://doi.org/10.1080/01419870.2011.559262

Perchoux, C., Chaix, B., Cummins, S., & Kestens, Y. (2013). Conceptualization and measurement of environmental exposure in epidemiology: Accounting for activity space related to daily mobility. *Health & Place, 21*, 86–93. https://doi.org/10.1016/j.healthplace.2013.01.005

PERI. (2021). *Air toxics at school*. Political Economy Research Institute. https://grconnect.com/tox100/schoolry2018/index.php?search=yes&school_name=&city=steubenville&state=&state_sum=

Peterle, G. (2019). Carto-fiction: Narrativising maps through creative writing. *Social & Cultural Geography, 20*(8), 1070–1093. https://doi.org/10.1080/14649365.2018.1428820

Petteway, R. J. (2016, April 24). Keeping it 100 for minority health month in steubenville. *Herald-Star*. https://www.heraldstaronline.com/opinion/local-columns/2016/04/guest-column-keeping-it-100-for-minority-health-month-in-steubenville/

Petteway, R. (2017). *Real limits of imaginary lines: A participatory activity space method for exploring intergenerational (dis)connections between 'place' and health*. 145th meeting of the American Public Health Association.

Petteway, R. (2018, May 7). The real limits of census tracts, and other boundaries. *Shelterforce, Spring 2018*. https://shelterforce.org/2018/05/07/the-real-limits-of-imaginary-lines/

Petteway, R. J. (2019). Intergenerational photovoice perspectives of place and health in public housing: Participatory coding, theming, and mapping in/of the "structure struggle". *Health & Place, 60*, 102229. https://doi.org/10.1016/j.healthplace.2019.102229

Petteway, R. (2021). Dreams of a beloved public health: Confronting white supremacy in our field. *Health Affairs Blog*. https://www.healthaffairs.org/do/10.1377/hblog20210204.432267/full/

Petteway, R. J., Mujahid, M., Allen, A., & Morello-Frosch, R. (2019a). The body language of place: A new method for mapping intergenerational "geographies of embodiment" in place-health research. *Social Science & Medicine, 223*, 51–63. https://doi.org/10.1016/j.socscimed.2019.01.027

Petteway, R., Mujahid, M., & Allen, A. (2019b). Understanding embodiment in place-health research: Approaches, limitations, and opportunities. *Journal of Urban Health, 96*(2), 289–299. https://doi.org/10.1007/s11524-018-00336-y

Pfister, D. S., & Godana, G. D. (2012). Deliberation technology. *Journal of Public Deliberation, 8*(1), 7.

PHNCI. (2020). *The 10 essential public health services*. The Public Health National Center for Innovations. https://phnci.org/uploads/resource-files/EPHS-English.pdf

Pruitt, S. L., Jeffe, D. B., Yan, Y., & Schootman, M. (2012). Reliability of perceived neighbourhood conditions and the effects of measurement error on self-rated health across urban and rural neighbourhoods. *Journal of Epidemiology and Community Health, 66*(4), 342–351. https://doi.org/10.1136/jech.2009.103325

Ruglis, J. (2011). Mapping the biopolitics of school dropout and youth resistance. *International Journal of Qualitative Studies in Education, 24*(5), 627–637. https://doi.org/10.1080/0951839 8.2011.600268

RWJF. (2020). *County health rankings*. County Health Rankings & Roadmaps. https://www.countyhealthrankings.org/explore-health-rankings

Schulz, A. J., Mentz, G., Lachance, L., Zenk, S. N., Johnson, J., Stokes, C., & Mandell, R. (2013). Do observed or perceived characteristics of the neighborhood environment mediate associations between neighborhood poverty and cumulative biological risk? *Health & Place, 24*, 147–156. https://doi.org/10.1016/j.healthplace.2013.09.005

Shareck, M., & Ellaway, A. (2011). Neighbourhood crime and smoking: The role of objective and perceived crime measures. *BMC Public Health, 11*(1), 930.

SHD. (2012). *2012 annual report*. City of Steubenville Health Department.

Smylie, J., Lofters, A., Firestone, M., & O'Campo, P. (2012). Population-based data and community empowerment. In P. O'Campo & J. R. Dunn (Eds.), *Rethinking social epidemiology* (pp. 67–92). Springer. https://doi.org/10.1007/978-94-007-2138-8_4

Solórzano, D. G., & Yosso, T. J. (2002). Critical race methodology: Counter-storytelling as an analytical framework for education research. *Qualitative Inquiry, 8*(1), 23–44. https://doi.org/10.1177/107780040200800103

Swords, J., Jeffries, M., East, H., & Messer, S. (2019). Mapping the city: Participatory mapping with young people. *Geography, 104*(3), 141–147. https://doi.org/10.1080/00167487.201 9.12094077

TCE. (n.d.). *Building healthy communities*. The California Endowment: Building Healthy Communities Initiative. Retrieved June 9, 2021, from https://www.buildinghealthycommunities.org/

Tilt, J. H., Unfried, T. M., & Roca, B. (2007). Using objective and subjective measures of neighborhood greenness and accessible destinations for understanding walking trips and BMI in Seattle, Washington. *American Journal of Health Promotion, 21*(4_suppl), 371–379.

Van Wart, S., Tsai, K., & Parikh, T. (2010). *Local ground: A paper-based toolkit for documenting local geospatial knowledge*. ACM symposium on Computing for Development (DEV).

Velasco, G., Faria, C., & Walenta, J. (2020). Imagining environmental justice "Across the street": Zine-making as creative feminist geographic method. *GeoHumanities, 6*(2), 347–370. https://doi.org/10.1080/2373566X.2020.1814161

Weden, M. M., Carpiano, R. M., & Robert, S. A. (2008). Subjective and objective neighborhood characteristics and adult health. *Social Science & Medicine, 66*(6), 1256–1270. https://doi.org/10.1016/j.socscimed.2007.11.041

Wen, M., Hawkley, L. C., & Cacioppo, J. T. (2006). Objective and perceived neighborhood environment, individual SES and psychosocial factors, and self-rated health: An analysis of older adults in Cook County, Illinois. *Social Science & Medicine, 63*(10), 2575–2590. https://doi.org/10.1016/j.socscimed.2006.06.025

Chapter 6
Toward Decolonizing Place-Health Research: Placemaking, Power, and the Production of "Place"-Health Knowledge

(Mis)Representation

Placemaking and Health: Putting Power on the Map

> One needs to reflect upon US history and its troubling legacy of 'placemaking' manifested in acts of displacement, removal, and containment. This history is long and horrible, from the forced movement of American Indians from their lands and their confinement to reservations, the Chinese Exclusion Act of 1882, the internment of Japanese Americans during World War II, to the urban redevelopment movement of the 1960s and 1970s that destroyed working poor and ethnic neighborhoods across American cities using the language of blight alongside bulldozers. (Bedoya, 2013)

It is impossible to address the myriad ways in which place affects health if we fail to unpack how "place" comes about to begin with and what maintains/reproduces it as site and source of health inequities. Most work to date, however, tends to de-place place-health relationships by not explicitly engaging the social, political, and economic practices/processes that fundamentally create (socio)spatial distributions of health opportunities and risks, i.e., ahistoric, apolitical, power-blind, cross-sectional examinations. We must be more explicit in accounting for and responding to the ways in which place—as more than geographic location—is actively made, unmade, and remade over time. That is, we must develop and center a critical awareness of *placemaking* processes and mechanisms. In outlining the *placescape* framework in Chap. 3, particularly as response to the ahistoric and power-blind proclivities of most place-health research, the goal was to articulate a mode for "seeing" place as relationally (re)produced by/through social and political processes, rendered "real" via, for example, the materiality of literal makings, unmakings, and remakings of people's communities. This perspective is almost entirely absent in my read of the place-health literature, particularly within public health. And when the term "placemaking" is used, it is without any explicit connection to power and broader sociospatial and political contexts/histories that (re)produce(d) spatial arrangements

R. J. Petteway, *Representation, Re-Presentation, and Resistance*, Global Perspectives on Health Geography, https://doi.org/10.1007/978-3-031-06141-7_6

of health risks and opportunities. Yet the field could benefit greatly—on empirical, conceptual, and translational fronts (including sociopolitical engagement)—from deeper engagement with more critical notions of placemaking. Here, I draw from the notion of "relational place-making" as articulated by Pierce et al. (2011) to help orient place-health work toward deeper engagement with considerations of power—as both predictive/productive, and a production, of place and its representations.

Pierce et al. (2011) define place-making as, "the set of social, political and material processes by which people iteratively create and recreate the experienced geographies in which they live" (p. 54). Starting from there, and merging literatures regarding notions of networked politics and networked place, they articulate a view of "relational place-making" as "the networked, political processes of place-framing" (p. 54). In doing so, they draw heavily from Massey's (2005) work conceptualizing place as "temporary constellations" consisting of "bundles" of individual space-time trajectories that "have purpose and meaning, but whose members may be reclaimed and repurposed into other configurations when viewed from other perspectives" (Pierce et al., 2011, p. 58). As Pierce et al. (2011) describe, this (re)configuring process—as production and making of place—is simultaneously structured by broader sociopolitical and material forces, and by networked social actors exercising their own agency, such that "places develop from pervasive structural forces that produce particular built environments and values" (p. 60). In this light, as germane to place-health research, the material aspects of place as (un)made by structural forces cannot be understood divorced from the sociorelational aspects of those carrying out, witnessing, remembering, and resisting the (un)makings—a position not too dissimilar from those articulated by Cummins et al. (2007), Bambra et al. (2019), and Pred (1984), for example. Appreciating this, I believe, entails at least 2 core considerations and (re)orientations for place-health research.

First, it entails more expressly accounting for those broader structural forces that shape material contexts and drive the spatial (re)sorting of people in relation to health risks/opportunities. As discussed in earlier chapters, especially Chaps. 3, 4, and 5, processes and consequences of placemaking as a sociopolitical and material reality can be readily discerned in "reading" people's placescapes, activity space maps, and geographies of embodiment. It is important, of course, to distinguish placemaking as I use it in this context—as a literal and critical process/practice of producing material contexts that reflect/inform sociospatial power relationships—from the use common in urban planning and arts communities that centers notions of place belonging and place attachment, often connected to notions of "creative placemaking" (Crisman, 2021; Madsen, 2018). This latter use, of course, is not entirely separate from the one I employ, as it too represents an aspect of relational place-making. But a mural on a building became a specific mural on a specific building in a specific neighborhood in a specific time period not by pure chance and love of art.

My focus in this section, for illustrative purposes, is on what might be considered more of the literal component of placemaking—"PlaceMaking" rooted in place-taking, if you will—structuring fundamental relations of/to space, property, and capital that undergird place-health contexts across communities and geographies—with attending political and social production components withstanding. Some

prominent examples here in the United States, as a settler-colonial state characterized by the racialization of space/place in/as the spatial sorting and organization of
opportunity/risk (de Souza Briggs, 2005; Edwards & Thomson, 2010; Glenn, 2015;
Harvey, 2004; Kent-Stoll, 2020; Lipsitz, 2011; Neely & Samura, 2011; Powell &
Bullard, 2007; Powell & Cardwell, 2013; Saito, 2014; Shabazz, 2015) are shown in
Table 6.1. The intention is not to expound upon these elements—and their interrelations—in any exhaustive sense. Other scholars have spoken to and will continue to
speak to this in greater depth and with greater care than I can do here (Brown, 2021;
Freund, 2010; Fullilove, 2004; Glotzer, 2020; Gonda, 2019; Loewen, 2018; Massey
& Denton, 1993; Nichols, 2019; Powell & Spencer, 2002; Rothstein, 2018;
Trounstine, 2018; Vale, 2013). Rather, the intent is to simply note why it is necessary to ground examinations of place-health relationships in the sociospatial histories and contemporary contexts of dispossession, repossession, surveillance,

Table 6.1 PlaceMaking examples in the United States

PlaceMaking mechanism/process	Example literature
Indian Removal Act of 1830	Foreman and Debo (1974), Perdue et al. (2008), and Saunt (2020)
Land Claims and Homesteading Acts (e.g. Pre-Emption Act, Oregon Donations Land Act of 1850, Bounty Land Act of 1850, Homestead Act of 1862; Morrill Act)	Nash (2019) and Nichols (2019)
Dawes Act of 1887	McDonnell (1991) and Otis (1973)
Alien Land & Racial Property Exclusion Laws (e.g. California Alien Land Acts, Oregon Black Exclusion Laws)	Castleman (1994), Lazarus (1988), and Villazor (2009)
Racially Restrictive Covenants	Gonda (2019), Jones-Correa (2000), and Majumdar (2006)
Redlining	Brown (2021), Massey and Denton (1993), and Rothstein (2018)
Realtors Codes of Ethics	Helper (1969) and Mcelderry (2001)
Partition Sales/Torrens Act	Gaither et al. (2019)
Urban Renewal	Fullilove (2004), Fullilove (2001), and Fullilove and Wallace (2011)
Metropolitan Fragmentation	Powell (2002) and Rusk (1993)
Racialized and Exclusionary Zoning	Brown (2021), Power (1983), and Wilson et al. (2008)
HOPE VI	Goetz (2011, 2013a, b), NHLP (2002), and Vale (2013)
Gentrification	Goetz (2011), Kent-Stoll (2020), Moskowitz (2018), Powell and Spencer (2002), and Smith (1982)

Selected examples of "PlaceMaking" mechanisms in the United States. These are not intended to
be exhaustive. Rather, they serve as some illustrative and concrete examples of how the "places"
in our place-health studies were/are materially (re)produced in ways that fundamentally structure
spatial (and racial) patterns and configurations of health risks and opportunities—all of which
remain enmeshed with/emergent from historic and present arrangements of spatial power vis a vis
land/property ownership and control

exclusion, and erasure as fundamentally productive of place as material and inextricably linked to place as a sociopolitical production.

Second, it entails acknowledging that how a place is represented, and who is doing the representing, is itself one of those "pervasive structural forces" (Pierce et al., 2011, p. 60), i.e., a placemaking mechanism. Here, Pierce and colleagues draw from Martin's (2003) notion of "place-frames," describing them as the "discursive, political understandings of the process of place production" (Pierce et al., 2011, p. 55), to outline an orientation to place as inherently political and thus a site of continual contestation. I suggest that what is being—or must be—contested is at least three-part: (1) what a "place" is, (2) how a "place" came about, and (3) narratives and representations of "place."

As discussed in Chap. 1 specifically and in various capacities throughout this book, place-health research has much room for improvement on the first matter. My concern in Chaps. 2, 3, 4 and 5, and more expressly in this chapter, is the latter two—including their interactions with each other and with the first item. In short, our failure to engage/unpack the second item not only condemns our work as perpetually detached and decontextualized, but all but guarantees that narratives about and representations of place—the third item—will remain ahistoric and blind to power, thereby functioning as a *de-placing* placemaking mechanism of productive/repressive power in relation to spatial knowledge(s). As articulated by Elwood and Leszczynski (2018), in drawing from the work of McKittrick (2016) and others, "theorizing social and spatial formations through histories of racial, colonial, gendered, and heteronormative domination is crucial: it makes these forms of domination legible and tractable for intervention" (p. 639). Centering these modes of spatial domination and exclusion, as both exercised and rendered visible through place (and representations thereof) should be of special interest within place-health research given the robust body of work documenting, for example, racialized spatial health inequities (André Hutson et al., 2012; Kotecki et al., 2019; Mehra et al., 2017; Merkin et al., 2009; Morello-Frosch & Jesdale, 2006; Morello-Frosch & Lopez, 2006; Williams & Collins, 2001). In the context of place-health research, failure to engage these notions of/as placemaking may very well foreclose the possibility of meaningful intervention and the development of place-health counternarratives.

As Allen et al. (2019) articulate, "relational place-making incorporates the selection not only of symbolic, discursive elements but also material, physical elements" (p. 1012). Placemaking must be understood as social, political, material, *and* symbolic/representational. These processes and mechanisms are interrelated and iterative, producing material realities and narratives/representations of place that inform how other expressions of spatial power and placemaking will be brought to bear on/for/against a place and its residents (see, e.g., Bloom et al., 2015; Goetz, 2013a; Vale, 2013). I think it is important to highlight the interplay between the material and the symbolic and to foreground the material aspect for a moment via Table 6.1. More so, I believe it is critical to more explicitly interrogate/explicate place-health knowledge production itself as placemaking—symbolic and representational with material consequences.

Re-Presentation and/as Resistance

Toward Decolonizing Place-Health Knowledges, Narratives, and Representations: Place-Health Research as Placemaking

Representation is important as a concept because it gives the impression of 'the truth'...
There are problems, too, when we do see ourselves but can barely recognize ourselves
through the representation. (Smith, 2013, p. 37)

Interrogating (Mis)Representation(s)

Accounting for power in placemaking within our research should allow for us to more thoroughly interrogate how place has been and continues to be (mis)represented, and whose place-health knowledge and narratives of place get mapped, so to speak. As I've suggested throughout this book and revisit below, this can allow for a process of place "rearticulation"—an opportunity to create counternarratives of place and place-health relationships therein. That is, engaging considerations of power and agency within discourse of place and health is the prerequisite first step toward decolonizing place-health research and knowledge production. Place-based health inequities are the product and (re)production of structural power dynamics—social, economic, and political. Yet, considerations of power are remarkably absent from most domains of place-health discourse. One critical yet perpetually overlooked domain is that of place-health *knowledge production* itself—the mechanisms, processes, and procedures for creating and *(de)valuing* knowledge(s) about place and health, and underlying/motivating epistemologies and norms therein. That is, questions of epistemic and data justice for place-health narratives and place (mis)representations. Here, there are a few core conceptual considerations that animate my thinking, which I believe can help orient place-health research/practice toward a praxis of resistance and decolonization.

As articulated in introducing the placescape in Chap. 3 (Tables 3.1 and 3.2), it is critical to account for power and agency in placemaking, and to understand that place-health research—and spatial narratives and representations therefrom—are a part of the placemaking process. We must be willing to engage questions regarding which and whose place knowledges are valued and deemed legitimate within place-health research and related policy discourse—that is, we must engage power-knowledge relations in place-health knowledge production. Unfortunately, most work has remained unapologetically apolitical and blind to power-knowledge relationships in pursuit of unabashedly positivist, reductionist, and largely cross-sectional, ahistoric, and decontextualized effect estimates as representations of place-health reality and "truth." Yet, as articulated by Foucault (1980), there is not a singular truth or even an "ensemble of truths which are to be discovered and accepted" (p. 132). Rather, there exists an "ensemble of rules according to which the true and the false are separated and specific effects of power attached to the true." He goes on to clarify that "it's not a matter of a battle 'on behalf' of the truth, but of a battle about the status of truth and the economic and political role it plays." The veil of objectivity and neutrality so characteristic of public health knowledge production accordingly must be interrogated constantly—as what constitutes/counts as knowledge, evidence, or "truth" is socially and politically contingent upon whom is doing the appraising and when. Moreover, not every form/expression of knowledge is even granted an opportunity to be appraised, i.e., epistemic oppression and erasure (Dotson, 2014). As noted by Martin (2003) and discussed above, place is inherently political and accordingly a site of continual contestation, and Bambra et al. (2019) make a strong case for political considerations regarding place and health specifically. Other scholars have done so regarding public health in general (Atwood et al., 1997; Hunter, 2016; Liverani et al., 2013; Morgan-Trimmer, 2014; Muntaner et al., 2012; Murphy & Fafard, 2012; Oliver, 2006; Smith, 2014). It seems rather clear: place and health—and knowledge production thereabout—are invariably (re)productions, and sites of, political contestation. From this position, in the context of place-health research and health geographies—and representations and narratives produced thereof—we should be asking ourselves a couple of simple questions: what are the rules regarding place-health knowledge production, and who gets to (re)set them?

Here, Foucault speaks of three domains—objects, ritual, and the privileged—that serve to set the rules, so to speak, for knowledge production and/as expressions of power. And by and large, place-health research represents rather poignantly the simultaneous presence of what Foucault (1980) describes as *repressive power* and *productive power* within power-knowledge relations. Each functions in tandem to create knowable "objects." Repressive power is observed, for example, in the manner in which people are prohibited from speaking for themselves, having to speak through the researcher, who strains to convert decades of context into an effect estimate with two decimal places—discretely knowable if you have $39.95 to pay for article access. This standard "ritual" of knowledge production—the discursive rules of practice, processes of engagement, and validation/valuation spaces—represses people's capacity to introduce themselves to, and represent themselves within, the knowledge world, and instead *produces* them into existence as scientific artifacts for the edification of elites. The credentialed researchers—whom Foucault refers to as "the privileged"—come to be viewed as "experts," usurping the space for place-health discourse using academic language to set a tone of manufactured "expertise"—an expertise made possible only through the exclusion and silencing of the people from which the knowledge underlying said expertise is derived. They use the people's place-health experiences—data—to craft stories about their lives and deaths—productions of place-health knowledge written, directed, and performed—in more ways than one—in an act of knowledge minstrelsy. The net effect is an overwhelmingly extractivist, reductionist, and (racial) capitalist knowledge production apparatus ("ritual") controlled disproportionately by scholars ("the privileged") whose social locations bear little resemblance to those of their study population ("objects").

In the context of public health, and specifically (social) epidemiology related to place-health research, these power-knowledge relations and discursive practices have arguably been defined by and coterminous with what Zuberi and Bonilla-Silva (2008, p. 17) refer to as "White logics": "a context in which White supremacy has defined the techniques and processes of reasoning about social facts." As they describe, "it assumes a historical posture that grants eternal objectivity to the views of elite Whites and condemns the views of non-Whites to perpetual subjectivity." This logic manifests in colorblind, apolitical, positivist methodological practices that center claims of detached neutrality and "objectivity", i.e., "White methods." These concepts in many ways help contextualize Moreton-Robinson's (2015) notion of "the White possessive"—generally, a "mode of rationalization… that is under-pinned by an excessive desire to invest in reproducing and reaffirming the nation-state's ownership, control, and domination" (p. xii). Importantly, Moreton-Robinson connects the notion of the White possessive to Foucault's notion of biopower (Chap. 9), specifically in regards to how racialization, for Foucault, became a primary mechanism through which disciplinary knowledges were deployed to regulate/sub-jugate populations and normalize (and invisiblize) whiteness—with important implications for/within discourses of sovereignty, representation, ownership, and knowledge production. For place-health research, this translates into colonizing, extractivist, and dispossessing research practices that center White logics and White ownership (of data/samples, i.e., bodies) and dominance (e.g., narrative control)—enacting epistemic violence and erasure via myriad social and material "significa-tions" of power and ownership, with bodies and communities as sites of place-health insertion/incursion.

In the context of place-health research, spatial place data and representations (e.g., maps) become objects and significations of both power and ownership, as do the norms regarding their access and use (see, e.g., paywalls, data user agreements and fees, discussion in CH5 regarding availability of/access to GIS data and maps). As articulated by Moreton-Robinson (2015, p. 50), "to be able to assert 'this is mine' requires a subject to internalize the idea that one has proprietary rights that are part of normative behavior, rules of interaction, and social engagement." For much place-health research, these rules (the "ritual"), as rooted in White logic and methods, center credentialed outsiders' ("the privileged") interpretations and repre-sentations of "place." Procedural and methodological norms—very much function-ing as technologies of biopower—preclude expressions of/actively mask community resident agency, discounting/devaluing their knowledge while simultaneously dis-possessing them of—and commodifying—their stories/experiences ("data"). Under these conditions, relationships between researchers and community represent not only the re-inscription of social hierarchy but the reification of place-health research as (racial) capitalist (re)colonization.

As Zuberi and Bonilla-Silva (2008) argue, researchers' social locations, e.g., race, gender, and class, represent "structural factors that partly shape researchers and their scientific gaze" (p. 18). In doing so, they note that all racialized groups have used and continue to use White logic and methods. Thus, the concern is both one of researcher's social location as well as researcher's epistemological orientation, hence the importance of considerations for decolonizing in order to move away from extractivist and dispossessing practices that underlly concerns of misrepresentation (Datta, 2018). While such considerations have been taken up within geography and urban planning work, including scholarship that informs my work here (Barry & Agyeman, 2020; Bonds & Inwood, 2016; de Leeuw & Hunt, 2018; Dorries et al., 2019; Launius & Boyce, 2021; Noxolo, 2017; Pulido, 2018; Radcliffe, 2017), there appears to be a pronounced lack of attention to these matters specifically in relation to place-heath relationships and health geographies.[1]

Importantly, I want to note here that I use decolonizing, in the context of place-health research, primarily in relation to knowledge, "map," and place narrative production as inherently colonizing acts—as the epistemic and procedural norms that animate place-health research materially and symbolically enact rituals of colonization, e.g., dispossession, expropriation, silencing, erasure, un-naming/renaming, commodification. In other words, here I'm concerned with the (re)colonization of place—and our lives/bodies—via knowledge production processes and subsequent (mis)representations and place-health narrative constructions. As noted above regarding placemaking, I see this as material and symbolic in nature, much in the way that Smith (2013) articulated materiality regarding decolonizing knowledges: "the production of knowledge, new knowledge and transformed 'old' knowledge, ideas about the nature of knowledge and the validity of specific forms of knowledge, became as much commodities of colonial exploitation as other natural resources" (p. 62). There are material and symbolic consequences of a place-health research that inserts itself into places for purposes of data extraction and commodification— profiting off "samples" (e.g., survey data, biological specimens) and (re)producing disembodied, decontextualized narratives of place. In other words, a decolonizing orientation offers a much-needed mode of resistance, rearticulation, and resurgence to counter the White supremacist, settler-colonial logics and methods that characterize the standard place-health research "ritual."

As discussed in various capacities throughout this book, traditional place-health knowledge production practices—in masking resident voice/agency, stripping away context, and foreclosing possibility of counternarrative—function to center and serve "outsider" narratives of place in a mold that settler-colonial, White logics can accommodate as legitimate. As illustrated in Chaps. 4 and 5 regarding the racial and spatial misrepresentation and erasure of Black residents' place knowledges and sense of "place," these practices tend toward enacting what might be considered a form of "urbicide"—a killing of the city (Campbell et al., 2007). While the notion of urbicide has largely been used to describe sociomaterial processes/consequences of deliberate (and racialized) disinvestment, devaluation, and infrastructural decomposition, dispossession, and erasure, I believe it captures both sociomaterial and symbolic aspects of the work presented in previous chapters. For example, sociomaterial in the context discussed in Chap. 3 regarding infrastructural decomposition of

---

[1] For example, a PubMed search using (decoloniz*[Title/Abstract]) AND ((place[Title/Abstract] OR neighborhood)[Title/Abstract]) finds less than 40 results—none of which are relevant/on topic

residents' housing units/community and the dispossession—unmaking and remaking—of public housing units for university dorms. And symbolic in the manner in which residents' place—at least as encountered and known by Black residents—is seemingly in a perpetual state of death/resurrection: killed by data omission and place misrepresentation/erasure, resurrected/alive in and through daily life and their embodied knowledge of place.

As articulated by McKittrick (2011), "the complexity of a black sense of place—questions of encounter, practices of resistance—can be, at least *conceptually*, swallowed up by the very death and decay that is bolstered by the hard empirical evidence of black geographic peril" (p. 951). For public health broadly, this is reflected in the propensity to center narratives of deficits and need and cannot be disconnected from concerns of spatial stigma within place-health research (Cairns, 2018; Graham et al., 2016; Halliday et al., 2020; Jewell, 2018; Tran et al., 2020). As the figures in Chaps. 3, 4, and 5 show, traditional data practices and mapping processes alone would have largely misrepresented—experientially and spatially—residents' encounters with and sense of place, allowing to emerge only that which would have served to reinforce existing dominant—and stigmatizing—place narratives. This not only represents an expression of repressive power and epistemic oppression, but in effect, an erasure of a sense of place not defined entirely by racialized narratives of deficits, need, and "risks," created/conjured by "the privileged" as rooted in White, colonial logics.

Such repression and erasure simultaneously presents as an expression of productive power—with residents' "place" produced into existence and rendered real for/by the gaze of "the privileged." This is how you can have, as noted in Chap. 5, for example, City administrators describe entire Black communities as "lacking a sense of place," without having had any conversations/interactions with residents—and no existing mechanism that would require such conversations/interactions or deem them of potential value. While this particular narrative was produced by City administrators and not researchers, I think practitioners and researchers alike would do well to reflect on Hunter et al.' (2016) concept of "black placemaking," framed as a practice that "attempts to counter the scholarship that contributes to the unrelenting negative portrayals of black neighborhoods without losing sight of the dialectical relationship between structure and agency, between domination and resistance." Whether related specifically to Black communities or not, there remains much room for improvement for place-health scholars to engage these dialectics within/as placemaking processes. And even more room for interrogating our role/complicity in domination within knowledge production processes. A core question then becomes, "how do we move toward a decolonized place-health research and discourse of re-presentation and resistance?"

A Praxis of Re-presentation + Resistance

As discussed previously, notions of placemaking must be centered within place-health research, and place-health knowledge production must be viewed as the critical placemaking mechanism that it is. We accordingly must continue to interrogate and recalibrate our "ritual"—to improve empirical, conceptual, procedural, methodological, and, importantly, representational aspects of our work. Decolonizing orientations, in combination/alignment with key conceptual groundings of critical, critical race, Black feminist, and participatory research traditions, can offer a rich scaffolding upon/from which place-health scholars can begin to reimagine and remake the architecture and archive of place-health research and knowledges. There are a few core conceptual groundings that inform my thinking here. The intention is not to expound upon them in any measure of detail, nor to provide an exhaustive synthesis; rather, the aim is simply to use this space to begin to bring these concepts and conversations more expressly into the fold of place-health discourse, particularly within public health.

Public health has a long and proud history of engaging Paulo Freire's notion of praxis: "reflection and action upon the world in order to change it" (2000 16; p. 51). Generally, praxis is understood as the unity of thought and action, of theory and practice, rooted in critical self-reflection in relation to everyday life and the social and political sources/contexts of oppression that shape it. In other words, it is a process of *becoming* critical and way of *being*. As articulated by Ledwith (Ledwith, 2016), "in this way, instead of theory becoming detached from action, it is woven together and deepened in action as we work together to co-create knowledge that comes from experience, rather than unquestioningly accepting knowledge that is detached from life" (p. 58). The fundamental element of praxis is that it is rooted in lived experience and that the knowledge of that lived experience cannot be possessed or "known" in the same manner (if at all) by someone who hasn't lived it. In the context of place-health research, those who actually experience the material and social conditions of place-based health inequities possess a unique knowledge of them and thus a unique power. This power, realized through praxis, is very much a "power from the margins," as described by hooks (hooks, 2000), and very much aligned with notions of situated and decolonized knowledges (Haraway, 1988; Smith, 2013). Together, praxis and *power from the margins* are encapsulated in what is referred to in critical race theory as "centering the margins" (Ford & Airhihenbuwa, 2010a). From a *public health critical race praxis* perspective, this facilitates honoring "voice"—that is, "prioritizing the perspectives of marginalized persons" (Ford & Airhihenbuwa, 2010b, p. 1394). This not only enables the (co)production and inclusion of new knowledges but also offers avenues toward "disciplinary self-critique"—understood as "the systematic examination by members of a discipline of its conventions and impacts on the broader society" (Ford & Airhihenbuwa, 2010b, p. 1394).

By engaging these principles, place-health research can take on a critical role as/in what Antonio Gramsci referred to as *rearticulation*. As described by Ledwith (Ledwith, 2016), rearticulation is understood "as the dismantling of hegemonic stories in order to create counternarratives that capture new possibilities for a more just and equal future" (p. 145). In direct reference to praxis as a precursor, Ledwith continues, "imagining a better way forward needs to be built on an understanding of what is going on in the present, and what led up to that in the past" (p. 145). Praxis requires a lived and critical awareness of one's own experiences in relation to power, thus rearticulation as action and an extension of praxis requires that the people experiencing/embodying realities of oppression retain narrative control—not "the privileged." In the context of place-health research, those who are experiencing place-health inequities are uniquely positioned to "challenge the hegemonic narratives of everyday life" (Ledwith, 2016, p. xi), here as constructed about their "place"—they can produce themselves, and their place, into existence.

The notion of counternarratives, of course, finds symmetry within critical race discourse via the notion of counterstorytelling (Delgado, 1989; Delgado et al., 2017; Kobayashi, 2005; Solórzano & Yosso, 2002), within decolonizing literature, most notably Smith's (2013) *Decolonizing Methods*, and in geography via notions of countermapping (Hunt & Stevenson, 2017; Kidd, 2019; Louis et al., 2012). As articulated by Delgado (1989), counterstories "can show that what we believe is ridiculous, self-serving, or cruel… can show us the way out of the trap of unjustified exclusion… can help us understand when it is time to reallocate power" (p. 2415). In doing so, he frames counterstories as both productive, and critically, *destructive*—that is, they can help us to (re)imagine and act to pursue just and anti-oppressive narratives, and also help us to interrogate and dismantle those serving to curate and incubate exclusion and oppression. What I've suggested here, and as I described in Chaps. 2, 3, 4 and 5, is that place-health research—as placemaking via discursive practices of place (mis)representation—has a propensity to function as curator and incubator. Moreover, to the extent that it fails to engage broader notions of placemaking—policies, practices, and processes that historically/presently make, unmake, and remake the sociomaterial contexts of the "place" being studied—it can become knowledge for knowledge's sake: materially relevant only in regards to dispossession, expropriation, and commodification (e.g., $39.95).

Place-health researchers—if indeed animated and motivated by a desire to mitigate place-related health inequities—need to consider that, just maybe, our standard "ritual" forecloses the possibility of anti-oppressive, antiracist, decolonized narratives and representations of place. And if so, how do we imagine our work can produce knowledges capable of speaking to/reallocating power as brought to bear and expressed within the places we do our work? How does masking resident agency and excluding their lived and embodied knowledges of place serve to advance equity? As I've detailed in previous chapters, this is a baseline errant on empirical and conceptual grounds. And it's especially errant in regards to representation and epistemic justice. As Harwood et al. (2018) note, "not only does looking at the everyday lived experiences of subordinated populations disrupt the colonizer's view of the world but it challenges the normativity of whiteness and white privilege and reveals interlocking systems of oppression" (p. 1246). Envisioning place-health research as a space in which to (co)develop counterstories—as resistance and rearticulation—can help us to better orient our work toward decolonizing and epistemic equity, which perhaps can render our work more relevant and action-able, e.g., vis a vis centering the narratives of residents as networked social actors and political constituents (as opposed to samples and specimens).

Ultimately, as has been argued elsewhere (Tuck & Yang, 2012), decolonizing is not only about countering epistemic violence and shifting power relations within knowledge production (here, about place and health), but also about *materially* shifting power relations that animate pervading structures/processes of capital and land ownership/acquisition, spatial dispossession and (re)possession, and place dominance and erasure as rooted in White supremacy and settler-colonialism. That is, ideally, conceptions and discourse of decolonizing, within place-health work or otherwise, should lead to "action in the streets" (Rivera Cusicanqu, 2014), which for some means a return of land—that is to say, *all* of the places. Here, I must acknowl-edge, in using "decolonizing" in the manner I have here—most directly concerning knowledge production (very much influenced by Smith, 2013)—that I have at least partially engaged in what Tuck & Yang (2012) refer to as the "colonial equivoca-tion" version of deploying decolonizing as a metaphor. Though as I have noted throughout, I view place-health research as a mode of placemaking, of both sym-bolic and material consequence for how place is (mis)represented, enacted, and acted upon. While I personally cannot imagine doing place-health research that is so impactful that it not only leads to action in the streets but a return of some measure of land, I do not foreclose that as a desire and possibility for others'/our collective engagements.

Alas, in the spirit of Fanon, decolonizing the mind—the logics and epistemolo-gies that animate the dominant "ritual" of place-health research—is a necessary first step. For place-health researchers, especially those with public health backgrounds, community-based participatory research (CBPR) represents perhaps the most developed orientation to collaborative knowledge production for social action. Seen as more of an orientation to research than a method or set of methods, CBPR is generally characterized by equitable, collaborative, and mutually beneficial engage-ment between outside researchers, community residents, and other local stakehold-ers in the research process (Israel et al., 1998, 2010; Minkler, 2010; Wallerstein & Duran, 2010). At its core are principles concerning equity, power, and inclusive notions of knowledge and expertise, with roots in Freirean praxis, feminist theory, and other participatory traditions (Wallerstein & Duran, 2017). CBPR differs from traditional public health research approaches, including those applied in most place-health work, by: (1) involving equitable participation and co-learning among study participants/co-researchers and academic partners, (2) building on community strengths, assets, knowledges, and expertise, (3) centering considerations for build-ing and amplifying participant/community power, agency, and local capacity build-ing to address the factors under study, and (4) balancing research and action.

Much place-health research has a focus on social and environmental processes and exposures that are difficult to elucidate, measure, and act upon. Grounding this work in CBPR affords opportunities to improve place-health science in this regard, particularly because CBPR approaches can: (1) enable the development of more refined and relevant research questions, (2) improve research design and implementation strategies, (3) improve data collection and analysis, (4) afford broader reach for dissemination, (5) provide an explicit and more direct link to knowledge translation and social action, and (6) increase local capacity to sustain research and change efforts (Balazs & Morello-Frosch, 2013; Cargo & Mercer, 2008; Cook, 2008; Horowitz et al., 2009; Minkler, 2005). As I've noted in previous chapters and have discussed elsewhere (Petteway et al., 2019), integration of place-health work with CBPR is an opportunity to leverage the practical and procedural translational advantages of much place-based research (e.g., often space-bound, locality- and/or jurisdictionally specific), while simultaneously capitalizing on the scientific and political translational advantages of harnessing place-based knowledge, insight, and expertise of the people whose lives unfold within the "place" being studied. In other words, it is very much suited for centering the margins and facilitating the production of place-health counternarratives. In this capacity, CBPR presents as an established and respected research orientation that might serve as a sort of container from within which place-health researchers can begin to move toward decolonizing their work toward symbolic/representational ends.

Moreover, the centrality of knowledge production for social action within CBPR aligns with, as I've suggested, the imperative to more expressly engage notions of placemaking, particularly in relation to local/regional sociopolitical mechanisms as implicated in the production of material conditions associated with place-based health inequities. My concern in this chapter has primarily been oriented around considerations of power in placemaking, of which I consider place-health knowledge production a key component. Ultimately, for place-health research, it's about power as exercised/reallocated to alter the social, political, and economic structures that materially influence sociospatial arrangements of health risks/opportunities. Here, place-health and health geography scholars may find value in engaging scholarship concerning the forms, uses, and representations of power within local community contexts in order to more explicitly ground our work in considerations of power for research translation (Blackwell et al., 2007; Gilmore, 2002; Pearce, 2013; Popay et al., 2020; Powell et al., 2020; Smylie et al., 2012)—that is, action in the streets.

On the Way To…

Of course, CBPR is not without its limitations and tensions (Chávez et al., 2008; Darroch & Giles, 2014; Israel et al., 2006; Janes, 2016; de Leeuw et al., 2012; Minkler, 2005). Most of the research reported on in this book was rooted in CBPR principles and informed by much of the conceptual groundings discussed here. Even so, engagement of these topics requires constant reflection and reflexivity, perhaps especially so when one—as a researcher—is showing up in (and with) community on a variable spectrum somewhere between invitation and incursion—and all with the possibility of having been imbued with or internalized colonial logics, regardless of our social locations. The cautions from Fanon (2005) (e.g., colonized intellectuals) and Smith (2013) (e.g., brief-case carrying Indigenous people) should not be discounted, much as Zuberi and Bonilla-Silva (2008) noted that White logics and methods have been and continue to be deployed by scholars of all racialized groups. As a Black, descendant-of-enslaved-Africans, White, seasonally light, cis, straight, able-bodied professor who was born in a redlined city, raised in sundown towns, grew up in a single-parent home and public housing on $18,000/year with two brothers in a city that served as the impetus for the US Environmental Protections Agency's early regulations regarding air pollution and is the first (and only) in my family to graduate from college, I "know" a few things about "place" and health. And I also know that I've trained at three of the most "elite" public universities in the world, each of which is located on unceded Indigenous lands (at least one via Morrill Act land grant), and each of which loves positivist White logics and methods like I love Comté cheese and Oregon pinot noir.

So, in writing this book—and this Chapter most specifically—I do not discount the prospect that aspects of my work—consciously or not—render me complicit, on some level, in the maintenance of White supremacist settler-colonial logics in the production of knowledge about place and health. Because ultimately, *all of the places* that are places of place-health work in this settler-colonial state must be repatriated to render decolonizing, fully, more than metaphor (Tuck & Yang, 2012). Here, I find great value in and suggest to other place-health scholars Dotson's (2018) piece, wherein she outlines Black feminist identity politics practice as "on the way to decolonization." As she describes, Black feminist identity politics:

> *is a practice that aids in retaining the possibility of decolonization in a settler colonial state by resisting the historical unknowing that facilitates settler futurity.* It is not itself decolonization in a settler colony, but rather it is 'on the way' to such decolonization by resisting forms of historical unknowing and plays at innocence that further settler futures and anti-Blackness in the guise of liberation. (p. 190)

The choice to use the word "toward" in the title of this chapter was similarly constituted, in that I recognize that decolonizing—in the manner that animates my concerns regarding place-health research here—entails much more than considerations of power in placemaking and knowledge production. Decolonizing the standard "ritual" of place-health research—with its rituals of dispossession, expropriation, silencing, erasure, and commodification—is, perhaps, "on the way to" decolonization. As detailed throughout this book, place remains a prominent focus of both public health research and practice, and, as discussed in Chap. 3, there has accordingly been—and will continue to be—much attention directed toward place-focused research and place-based interventions. It should concern us all that this work has been referred to as the "new frontier" of public health (Amaro, 2014), as I cannot think of a more fundamentally errant and symbolically violent frame through which to reify place-health work as (re)colonization, dispossession, and erasure. The intention here has been to explicitly name this and call us in, so that we can re-up and reimagine what it means to "represent" place in our work, and be reminded what's at stake in our rituals of "place" misrepresentation.

Conclusion

In sum, our place-health work must account for power, not only in the social and material making, unmaking, and remaking of place but also in the representations and narratives of place—within in which, of course, our research is included. As illustrated in Chaps. 3, 4, and 5, considerations of power for/in placemaking must be at the center of any discourse regarding the spatial distribution of place-based health risks/opportunities and patterns therein. People's placescapes are not natural outgrowths of objective, agnostic geographic and topographic landscapes—they are irrigated, curated, and litigated material and symbolic productions of spatialized power. Thus, as suggested in Chap. 5, what ultimately lands upon our bodies is a map of power. I've suggested throughout this book that participatory, intergenerational approaches rooted in principles/practices of community-based participatory research—and informed by critical, critical race, and Black feminist theoretical orientations—can facilitate our movement toward a more inclusive, epistemically just place-health field. Place-health researchers would do well to recalibrate our methodological and procedural norms toward more inclusive approaches to place-health knowledge production that afford space for place-health counternarratives to refine place knowledge(s) and/or resist forces of symbolic and epistemic erasure. In doing so, we can pursue a more thorough engagement with notions of power in placemaking and, perhaps, move the field away from logics and methods of dispossession and misrepresentation, and closer to praxes and practices of decolonization and resistance.

References

Allen, D., Lawhon, M., & Pierce, J. (2019). Placing race: On the resonance of place with black geographies. *Progress in Human Geography, 43*(6), 1001–1019. https://doi.org/10.1177/0309132518803775

Amaro, H. (2014). The action is upstream: Place-based approaches for achieving population health and health equity. *American Journal of Public Health, 104*(6), 964.

André Hutson, M., Kaplan, G. A., Ranjit, N., & Mujahid, M. S. (2012). Metropolitan fragmentation and health disparities: Is there a link?: Metropolitan fragmentation and health disparities. *Milbank Quarterly, 90*(1), 187–207. https://doi.org/10.1111/j.1468-0009.2011.00659.x

Atwood, K., Colditz, G. A., & Kawachi, I. (1997). From public health science to prevention policy: Placing science in its social and political contexts. *American Journal of Public Health, 87*(10), 1603–1606. https://doi.org/10.2105/AJPH.87.10.1603

Balazs, C. L., & Morello-Frosch, R. (2013). The three Rs: How community-based participatory research strengthens the rigor, relevance, and reach of science. *Environmental Justice, 6*(1), 9–16. https://doi.org/10.1089/env.2012.0017

Bambra, C., Smith, K. E., & Pearce, J. (2019). Scaling up: The politics of health and place. *Social Science & Medicine, 232*, 36–42. https://doi.org/10.1016/j.socscimed.2019.04.036

Barry, J., & Agyeman, J. (2020). On belonging and becoming in the settler-colonial city: Co-produced futurities, placemaking, and urban planning in the United States. *Journal of Race, Ethnicity and the City, 1*(1–2), 22–41. https://doi.org/10.1080/26884674.2020.1793703

Bedoya, R. (2013). Placemaking and the politics of belonging and dis-belonging. *Grantmakers in the Arts*. https://www.giarts.org/article/placemaking-and-politics-belonging-and-dis-belonging

Blackwell, A. G., Blakely, E. J., Bositis, D. A., Cashin, S., Darden, J. T., Grigsby J. E. III., Johnson, G. S., Powell, J. A., Stoll, M. A., Torres, A., & Wright, B. (2007). *The black Metropolis in the twenty-first century: Race, power, and politics of place* (R. D. Bullard, Ed.). Rowman & Littlefield Publishers.

Bloom, N. D., Umbach, F., & Vale, L. J. (Eds.). (2015). *Public housing myths: Perception, reality, and social policy* (Illustrated edn.). Cornell University Press.

Bonds, A., & Inwood, J. (2016). Beyond white privilege: Geographies of white supremacy and settler colonialism. *Progress in Human Geography, 40*(6), 715–733. https://doi.org/10.1177/0309132515613166

Brown, L. T. (2021). *The black butterfly: The harmful politics of race and space in America*. Johns Hopkins University Press. https://jhupbooks.press.jhu.edu/title/black-butterfly

Cairns, K. (2018). Youth, temporality, and territorial stigma: Finding good in Camden, New Jersey. *Antipode, 50*(5), 1224–1243. https://doi.org/10.1111/anti.12407

Campbell, D., Graham, S., & Monk, D. B. (2007). Introduction to Urbicide: The killing of cities? *Theory & Event, 10*(2). https://doi.org/10.1353/tae.2007.0055

Cargo, M., & Mercer, S. L. (2008). The value and challenges of participatory research: Strengthening its practice. *Annual Review of Public Health, 29*(1), 325–350. https://doi.org/10.1146/annurev.publhealth.29.091307.083824

Castleman, B. A. (1994). California's alien land laws. *Western Legal History: The Journal of the Ninth Judicial Circuit Historical Society, 7*, 25.

Chávez, V., Duran, B., Baker, Q. E., Avila, M. M., & Wallerstein, N. (2008). The dance of race and privilege in CBPR. In M. Minkler & N. Wallerstein (Eds.), *Community-based participatory research for health: From process to outcomes* (2nd ed., pp. 91–106). Jossey-Bass.

Cook, W. K. (2008). Integrating research and action: A systematic review of community-based participatory research to address health disparities in environmental and occupational health in the USA. *Journal of Epidemiology & Community Health, 62*(8), 668–676. https://doi.org/10.1136/jech.2007.067645

Crisman, J. J. (2021). Evaluating values in creative placemaking: The arts as community development in the NEA's our town program. *Journal of Urban Affairs, 1*–19. https://doi.org/10.1080/07352166.2021.1890607

Cummins, S., Curtis, S., Diez-Roux, A. V., & Macintyre, S. (2007). Understanding and representing 'place' in health research: A relational approach. *Social Science & Medicine, 65*(9), 1825–1838. https://doi.org/10.1016/j.socscimed.2007.05.036

Darroch, F., & Giles, A. (2014). Decolonizing health research: Community-based participatory research and postcolonial feminist theory. *Canadian Journal of Action Research, 15*(3), 22–36.

Datta, R. (2018). Decolonizing both researcher and research and its effectiveness in Indigenous research. *Research Ethics, 14*(2), 1–24. https://doi.org/10.1177/1747016117733296

de Leeuw, S., & Hunt, S. (2018). Unsettling decolonizing geographies. *Geography Compass, 12*(7), e12376. https://doi.org/10.1111/gec3.12376

de Leeuw, S., Cameron, E. S., & Greenwood, M. L. (2012). Participatory and community-based research, indigenous geographies, and the spaces of friendship: A critical engagement: A critical engagement. *The Canadian Geographer/Le Géographe Canadien, 56*(2), 180–194. https://doi.org/10.1111/j.1541-0064.2012.00434.x

de Souza Briggs, X. (2005). *The geography of opportunity: Race and housing choice in Metropolitan America*. Brookings Institution Press. https://www.jstor.org/stable/10.7864/j.ctt1gpccgb

Delgado, R. (1989). Storytelling for oppositionists and others: A plea for narrative. *Michigan Law Review, 87*(8), 2411–2441. https://doi.org/10.2307/1289308

Delgado, R., Stefancic, J., & Harris, A. (2017). *Critical race theory* (3rd ed.). NYU Press.

Dorries, H., Hugill, D., & Tomiak, J. (2019). Racial capitalism and the production of settler colonial cities. *Geoforum, 132*, S001671851930226X. https://doi.org/10.1016/j.geoforum.2019.07.016

Dotson, K. (2018). On the way to decolonization in a settler colony: Re-introducing Black feminist identity politics. *AlterNative: An International Journal of Indigenous Peoples, 14*(3), 190–199. https://doi.org/10.1177/1177180118783301

Edwards, F. L., & Thomson, G. B. (2010). The legal creation of raced space: The subtle and ongoing discrimination created through Jim Crow Laws. *Berkeley Journal of African-American Law and Policy, 12*, 145.

Elwood, S., & Leszczynski, A. (2018). Feminist digital geographies. *Gender, Place & Culture, 25*(5), 629–644. https://doi.org/10.1080/0966369X.2018.1465396

Fanon, F., Sartre, J.-P., & Bhabha, H. K. (2005). *The Wretched of the Earth* (R. Philcox, Trans.; Reprint edn.). Grove Press.

Ford, C. L., & Airhihenbuwa, C. O. (2010a). Critical race theory, race equity, and public health: Toward antiracism praxis. *American Journal of Public Health, 100*(S1), S30–S35. https://doi.org/10.2105/AJPH.2009.171058

Ford, C. L., & Airhihenbuwa, C. O. (2010b). The public health critical race methodology: Praxis for antiracism research. *Social Science & Medicine, 71*(8), 1390–1398. https://doi.org/10.1016/j.socscimed.2010.07.030

Foreman, G., & Debo, A. (1974). *Indian removal: The emigration of the five civilized tribes of Indians*. (9th Bison Printing edn.). University of Oklahoma Press.

Foucault, M. (1980). In C. Gordon (Ed.), *Power/knowledge: Selected interviews and other writings, 1972–1977*. Pantheon Books.

Freire, P. (2000). *Pedagogy of the oppressed (30th anniversary ed)*. Continuum.

Freund, D. M. P. (2010). *Colored property: State policy and white racial politics in suburban America* (Illustrated edn.). University of Chicago Press.

Fullilove, M. T. (2001). Root shock: The consequences of African American dispossession. *Journal of Urban Health, 78*(1), 72–80.

Fullilove, M. (2004). *Root shock: How tearing up City neighborhoods hurts America, and what we can do about it*. New Village Press. https://nyupress.org/9781613320198/root-shock

Fullilove, M. T., & Wallace, R. (2011). Serial forced displacement in American cities, 1916–2010. *Journal of Urban Health, 88*(3), 381–389. https://doi.org/10.1007/s11524-011-9585-2

Gaither, C. J., Carpenter, A., Lloyd McCurty, T., & Toering, S. (2019). *Heirs' property and land fractionation: Fostering stable ownership to prevent land loss and abandonment* (SRS-GTR-244; p. SRS-GTR-244). U.S. Department of Agriculture, Forest Service, Southern Research Station. https://doi.org/10.2737/SRS-GTR-244

Gilmore, R. W. (2002). Fatal couplings of Power and difference: Notes on racism and geography. *The Professional Geographer, 54*(1), 15–24. https://doi.org/10.1111/0033-0124.00310

Glenn, E. N. (2015). Settler colonialism as structure: A framework for comparative studies of U.S. race and gender formation. *Sociology of Race and Ethnicity, 1*(1), 52–72. https://doi.org/10.1177/2332649214560440

Glotzer, P. (2020). *How the suburbs were segregated: Developers and the business of exclusionary housing, 1890–1960*. Columbia University Press.

Goetz, E. (2011). Gentrification in black and white: The racial impact of public housing demolition in American cities. *Urban Studies, 48*(8), 1581–1604. https://doi.org/10.1177/0042098010375323

Goetz, E. (2013a). *New Deal ruins: Race, economic justice, and public housing policy*. Cornell University Press. https://www.cornellpress.cornell.edu/book/9780801478284/new-deal-ruins/

Goetz, E. (2013b). The audacity of HOPE VI: Discourse and the dismantling of public housing. *Cities, 35*, 342–348.

Gonda, J. (2019). *Unjust deeds: The restrictive covenant cases and the making of the civil rights movement*. UNC Press. https://uncpress.org/book/9781469654812/unjust-deeds/

Graham, L. F., Padilla, M. B., Lopez, W. D., Stern, A. M., Peterson, J., & Keene, D. E. (2016). Spatial stigma and health in postindustrial Detroit. *International Quarterly of Community Health Education, 36*(2), 105–113. https://doi.org/10.1177/0272684X15627800

Halliday, E., Popay, J., Anderson de Cuevas, R., & Wheeler, P. (2020). The elephant in the room? Why spatial stigma does not receive the public health attention it deserves. *Journal of Public Health (Oxford, England), 42*(1), 38–43. https://doi.org/10.1093/pubmed/fdy214

Haraway, D. (1988). Situated knowledges: The science question in feminism and the privilege of partial perspective. *Feminist Studies, 14*(3), 575–599. https://doi.org/10.2307/3178066

Harvey, D. (2004). The "New" imperialism: Accumulation by dispossession. *Socialist Register, 40*. https://socialistregister.com/index.php/srv/article/view/5811

Harwood, S. A., Mendenhall, R., Lee, S. S., Riopelle, C., & Huntt, M. B. (2018). Everyday racism in integrated spaces: Mapping the experiences of students of color at a diversifying predominantly white institution. *Annals of the American Association of Geographers, 108*(5), 1245–1259. https://doi.org/10.1080/24694452.2017.1419122

Helper, R. (1969). *Racial policies and practices of real estate brokers* (NED-New edn.). University of Minnesota Press. https://www.jstor.org/stable/10.5749/j.ctttjd9

hooks, bell. (2000). *Feminist theory: From margin to center*. Pluto Press.

Horowitz, C. R., Robinson, M., & Seifer, S. (2009). Community-based participatory research from the margin to the mainstream: Are researchers prepared? *Circulation, 119*(19), 2633–2642. https://doi.org/10.1161/CIRCULATIONAHA.107.729863

Hunt, D., & Stevenson, S. A. (2017). Decolonizing geographies of power: Indigenous digital counter-mapping practices on turtle Island. *Settler Colonial Studies, 7*(3), 372–392. https://doi.org/10.1080/2201473X.2016.1186311

Hunter, E. L. (2016). Politics and public health—Engaging the third rail. *Journal of Public Health Management and Practice, 22*(5), 436–441. https://doi.org/10.1097/PHH.0000000000000446

Hunter, M. A., Pattillo, M., Robinson, Z. F., & Taylor, K.-Y. (2016). Black Placemaking: Celebration, play, and poetry. *Theory, Culture & Society, 33*(7–8), 31–56. https://doi.org/10.1177/0263276416635259

Israel, B. A., Schulz, A. J., Parker, E. A., & Becker, A. B. (1998). Review of community-based research: Assessing partnership approaches to improve public health. *Annual Review of Public Health, 19*(1), 173–202.

Israel, B. A., Krieger, J., Vlahov, D., Ciske, S., Foley, M., Fortin, P., Guzman, J. R., Lichtenstein, R., McGranaghan, R., Palermo, A., & Tang, G. (2006). Challenges and facilitating factors in sustaining community-based participatory research partnerships: Lessons learned from the Detroit, New York City and Seattle urban research centers. *Journal of Urban Health, 83*(6), 1022–1040. https://doi.org/10.1007/s11524-006-9110-1

Israel, B. A., Coombe, C. M., Cheezum, R. R., Schulz, A. J., McGranaghan, R. J., Lichtenstein, R., Reyes, A. G., Clement, J., & Burris, A. (2010). Community-based participatory research: A capacity-building approach for policy advocacy aimed at eliminating health disparities. *American Journal of Public Health, 100*(11), 2094–2102. https://doi.org/10.2105/AJPH.2009.170506

Janes, J. E. (2016). Democratic encounters? Epistemic privilege, power, and community-based participatory action research. *Action Research, 14*(1), 72–87. https://doi.org/10.1177/1476750315579129

Jewell, J. O. (2018). 'An injurious effect on the neighbourhood': Narratives of Neighbourhood decline and racialised class identities in late nineteenth-century San Francisco. *Immigrants & Minorities, 36*(1), 1–19. https://doi.org/10.1080/02619288.2017.1355734

Jones-Correa, M. (2000). The origins and diffusion of racial restrictive covenants. *Political Science Quarterly, 115*(4), 541–568. https://doi.org/10.2307/2657609

Kent-Stoll, P. (2020). The racial and colonial dimensions of gentrification. *Sociology Compass*. https://doi.org/10.1111/soc4.12838

Kidd, D. (2019). Extra-activism: Counter-mapping and data justice. *Information, Communication & Society, 22*(7), 954–970. https://doi.org/10.1080/1369118X.2019.1581243

Kobayashi, A. (2005). Anti-racist feminism in geography: An agenda for social action. In *A companion to feminist geography* (pp. 32–40). Wiley. https://doi.org/10.1002/9780470996898.ch3

Kotecki, J. A., Gennuso, K. P., Givens, M. L., & Kindig, D. A. (2019). Separate and sick: Residential segregation and the health of children and youth in metropolitan statistical areas. *Journal of Urban Health: Bulletin of the New York Academy of Medicine, 96*(2), 149–158. https://doi.org/10.1007/s11524-018-00330-4

Launius, S., & Boyce, G. A. (2021). More than metaphor: Settler colonialism, frontier logic, and the continuities of racialized dispossession in a southwest U.S. City. *Annals of the American*

Association of Geographers, 111(1), 157–174. https://doi.org/10.1080/24694452.2020.175094 0

Lazarus, M. L. I. (1988). An historical analysis of alien land law: Washington Territory & (and) State 1853–1889. *University of Puget Sound Law Review, 12*, 197.

Ledwith, M. (2016). *Community development in action: Putting Freire into practice*. Policy Press. http://stats.lib.pdx.edu/proxy.php?url=http://search.ebscohost.com/login.aspx?direct=true&db=nlebk&AN=1573495&site=ehost-live

Lipsitz, G. (2011). *How racism takes place*. Temple University Press.

Liverani, M., Hawkins, B., & Parkhurst, J. O. (2013). Political and institutional influences on the use of evidence in public health policy. A systematic review. *PLoS One, 8*(10), e77404. https://doi.org/10.1371/journal.pone.0077404

Loewen, J. W. (2018). *Sundown towns: A hidden dimension of American racism* (Illustrated edn.). The New Press.

Louis, R. P., Johnson, J. T., & Pramono, A. H. (2012). Introduction: Indigenous cartographies and counter-mapping. *Cartographica: The International Journal for Geographic Information and Geovisualization, 47*(2), 77–79. https://doi.org/10.3138/carto.47.2.77

Madsen, W. (2018). Evaluation and creative placemaking: Using a critical realist model to explore the complexity. *Journal of Applied Arts & Health, 9*(3), 411–422. https://doi.org/10.1386/jaah.9.3.411_1

Majumdar, R. (2006). Racially restrictive covenants in the state of Washington: A primer for practitioners. *Seattle University Law Review, 30*, 23.

Martin, D. G. (2003). "Place-framing" as place-making: Constituting a neighborhood for organizing and activism. *Annals of the Association of American Geographers, 93*(3), 730–750. https://doi.org/10.1111/1467-8306.9303011

Massey, D. B. (2005). *For space* (1st ed.). SAGE Publications Ltd.

Massey, D. S., & Denton, N. A. (1993). *American apartheid: Segregation and the making of the underclass* (Later Printing edn.). Harvard University Press.

McDonnell, J. A. (1991). *The dispossession of the American Indian, 1887–1934* (Later Printing edn.). Indiana University Press.

Mcelderry, S. (2001). Building a West Coast Ghetto: African-American Housing in Portland, 1910–1960. *The Pacific Northwest Quarterly, 92*(3), 137–148.

McKittrick, K. (2011). On plantations, prisons, and a black sense of place. *Social & Cultural Geography, 12*(8), 947–963. https://doi.org/10.1080/14649365.2011.624280

McKittrick, K. (2016). Diachronic loops/deadweight tonnage/bad made measure. *Cultural Geographies, 23*(1), 3–18. https://doi.org/10.1177/1474474015612716

Mehra, R., Boyd, L. M., & Ickovics, J. R. (2017). Racial residential segregation and adverse birth outcomes: A systematic review and meta-analysis. *Social Science & Medicine, 191*, 237–250. https://doi.org/10.1016/j.socscimed.2017.09.018

Merkin, S. S., Basurto-Dávila, R., Karlamangla, A., Bird, C. E., Lurie, N., Escarce, J., & Seeman, T. (2009). Neighborhoods and cumulative biological risk profiles by race/ethnicity in a National Sample of U.S. Adults: NHANES III. *Annals of Epidemiology, 19*(3), 194–201. https://doi.org/10.1016/j.annepidem.2008.12.006

Minkler, M. (2005). Community-based research partnerships: Challenges and opportunities. *Journal of Urban Health: Bulletin of the New York Academy of Medicine, 82*(2_suppl_2), ii3–ii12. https://doi.org/10.1093/jurban/jti034

Minkler, M. (2010). Linking science and policy through community-based participatory research to study and address health disparities. *American Journal of Public Health, 100*(S1), S81–S87.

Morello-Frosch, R., & Jesdale, B. M. (2006). Separate and unequal: Residential segregation and estimated cancer risks associated with ambient air toxics in U.S. Metropolitan Areas. *Environmental Health Perspectives, 114*(3), 386–393. https://doi.org/10.1289/ehp.8500

Morello-Frosch, R., & Lopez, R. (2006). The riskscape and the color line: Examining the role of segregation in environmental health disparities. *Environmental Research, 102*(2), 181–196. https://doi.org/10.1016/j.envres.2006.05.007

Moreton-Robinson, A. (2015). *The white possessive: Property, power, and indigenous sovereignty*. University of Minnesota Press. https://www.upress.umn.edu/book-division/books/the-white-possessive

Morgan-Trimmer, S. (2014). Policy is political; our ideas about knowledge translation must be too. *Journal of Epidemiology and Community Health, 68*(11), 1010–1011. https://doi.org/10.1136/jech-2014-203820

Moskowitz, P. E. (2018). *How to kill a city: Gentrification, inequality, and the fight for the neighborhood* (Reprint edn.). Bold Type Books.

Muntaner, C., Chung, H., Murphy, K., & Ng, E. (2012). Barriers to knowledge production, knowledge translation, and urban health policy change: Ideological, economic, and political considerations. *Journal of Urban Health, 89*(6), 915–924. https://doi.org/10.1007/s11524-012-9699-1

Murphy, K., & Fafard, P. (2012). Knowledge translation and social epidemiology: Taking Power, politics and values seriously. In P. O'Campo & J. R. Dunn (Eds.), *Rethinking social epidemiology* (pp. 267–283). Springer. https://doi.org/10.1007/978-94-007-2138-8_13

Nash, M. A. (2019). Entangled pasts: Land-Grant colleges and American Indian dispossession. *History of Education Quarterly, 59*(4), 437–467. https://doi.org/10.1017/heq.2019.31

Neely, B., & Samura, M. (2011). Social geographies of race: Connecting race and space. *Ethnic and Racial Studies, 34*(11), 1933–1952. https://doi.org/10.1080/01419870.2011.559262

NHLP. (2002). *False HOPE: A critical assessment of the HOPE VI public housing redevelopment program*. National Housing Law Project.

Nichols, R. (2019). *Theft is property!: Dispossession and critical theory*. Duke University Press Books.

Noxolo, P. (2017). Introduction: Decolonising geographical knowledge in a colonised and re-colonising postcolonial world. *Area, 49*(3), 317–319. https://doi.org/10.1111/area.12370

Oliver, T. R. (2006). The politics of public health policy. *Annual Review of Public Health, 27*(1), 195–233. https://doi.org/10.1146/annurev.publhealth.25.101802.123126

Otis, D. S. (1973). *The Dawes act and the allotment of Indian lands*. Oklahoma University Press. https://www.oupress.com/books/14187239/the-dawes-act-and-the-allotment-of-indian-lan

Pearce, J. (2013). Power and the twenty-first century activist: From the neighbourhood to the square. *Development and Change, 44*(3), 639–663. https://doi.org/10.1111/dech.12035

Perdue, T., Green, M., & Calloway, C. (2008). *The Cherokee Nation and the trail of tears* (Illustrated edn.). Penguin Books.

Petteway, R., Mujahid, M., Allen, A., & Morello-Frosch, R. (2019). Towards a people's social epidemiology: Envisioning a more inclusive and equitable future for social Epi research and practice in the 21st century. *International Journal of Environmental Research and Public Health, 16*(20), 3983. https://doi.org/10.3390/ijerph16203983

Pierce, J., Martin, D. G., & Murphy, J. T. (2011). Relational place-making: The networked politics of place. *Transactions of the Institute of British Geographers, 36*(1), 54–70. https://doi.org/10.1111/j.1475-5661.2010.00411.x

Popay, J., Whitehead, M., Ponsford, R., Egan, M., & Mead, R. (2020). Power, control, communities and health inequalities I: Theories, concepts and analytical frameworks. *Health Promotion International, 36*, daaa133. https://doi.org/10.1093/heapro/daaa133

Powell, J. (2002). *Sprawl, fragmentation, and the persistence of racial inequality: Limiting civil rights by fragmenting space*. Urban Institute. https://trid.trb.org/view/690208

Powell, J. A., & Bullard, R. (2007). Structural racism and spatial Jim crow. In *The black Metropolis in the twenty-first century: Race, power, and the politics of place* (p. 41). Rowman & Littlefield Publishers.

Powell, J. A., & Cardwell, K. (2013). Homeownership, wealth, and the production of racialized space. *Joint Center for Housing Studies Harvard University*, Article IR. https://lawcat.berkeley.edu/record/1125958

Powell, J. A., & Spencer, M. L. (2002). Giving them the old one-two: Gentrification and the KO of impoverished urban dwellers of color. *Howard Law Journal, 46*, 433.

Powell, K., Barnes, A., Anderson de Cuevas, R., Bambra, C., Halliday, E., Lewis, S., McGill, R., Orton, L., Ponsford, R., Salway, S., Townsend, A., Whitehead, M., & Popay, J. (2020). Power, control, communities and health inequalities III: Participatory spaces—An English case. *Health Promotion International*, daaa059. https://doi.org/10.1093/heapro/daaa059

Power, G. (1983). Apartheid Baltimore style: The residential segregation ordinances of 1910-1913. *Maryland Law Review, 42*(2), 43.

Pred, A. (1984). Place as historically contingent process: Structuration and the time-geography of becoming places. *Annals of the Association of American Geographers, 74*(2), 279–297. https://doi.org/10.1111/j.1467-8306.1984.tb01453.x

Pulido, L. (2018). Geographies of race and ethnicity III: Settler colonialism and non-native people of color. *Progress in Human Geography, 42*(2), 309–318. https://doi.org/10.1177/0309132516686011

Radcliffe, S. A. (2017). Decolonising geographical knowledges. *Transactions of the Institute of British Geographers, 42*(3), 329–333. https://doi.org/10.1111/tran.12195

Rivera Cusicanqu, S. (2014). *The Potosí principle: Another view of totality*. The Hemispheric Institute. https://hemisphericinstitute.org/en/emisferica-11-1-decolonial-gesture/11-1-essays/the-potosi-principle-another-view-of-totality.html

Rothstein, R. (2018). *The color of law: A forgotten history of how our government segregated America* (Reprint edn.). Liveright.

Rusk, M. D. (1993). *Cities without suburbs* (1st ptg. edn.). Woodrow Wilson Center Press.

Saito, N. T. (2014). Tales of color and colonialism: Racial realism and settler colonial theory. *Florida A & M University Law Review, 10*(1), 109.

Saunt, C. (2020). *Unworthy Republic: The dispossession of native Americans and the road to Indian territory* (Illustrated edn.). W. W. Norton & Company.

Shabazz, R. (2015). *Spatializing blackness: Architectures of confinement and black masculinity in Chicago*. University of Illinois Press.

Smith, N. (1982). Gentrification and uneven development. *Economic Geography, 58*(2), 139. https://doi.org/10.2307/143793

Smith, P. L. T. (2013). *Decolonizing methodologies: Research and indigenous peoples*. Zed Books Ltd.

Smith, K. E. (2014). The politics of ideas: The complex interplay of health inequalities research and policy. *Science and Public Policy, 41*(5), 561–574. https://doi.org/10.1093/scipol/sct085

Smylie, J., Lofters, A., Firestone, M., & O'Campo, P. (2012). Population-based data and community empowerment. In P. O'Campo & J. R. Dunn (Eds.), *Rethinking social epidemiology* (pp. 67–92). Springer. https://doi.org/10.1007/978-94-007-2138-8_4

Solórzano, D. G., & Yosso, T. J. (2002). Critical race methodology: Counter-storytelling as an analytical framework for education research. *Qualitative Inquiry, 8*(1), 23–44. https://doi.org/10.1177/107780040200800103

Tran, E., Blankenship, K., Whittaker, S., Rosenberg, A., Schlesinger, P., Kershaw, T., & Keene, D. (2020). My neighborhood has a good reputation: Associations between spatial stigma and health. *Health & Place, 64*, 102392. https://doi.org/10.1016/j.healthplace.2020.102392

Trounstine, J. (2018). *Segregation by design: Local politics and inequality in American cities*. Cambridge University Press. https://doi.org/10.1017/9781108555722

Tuck, E., & Yang, K. W. (2012). Decolonization is not a metaphor. *Decolonization: Indigeneity, Education & Society, 1*(1), Article 1. https://jps.library.utoronto.ca/index.php/des/article/view/18630

Vale, L. J. (2013). *Purging the poorest: Public housing and the design politics of twice-cleared communities* (Illustrated edn.). University of Chicago Press.

Villazor, R. C. (2009). Rediscovering Oyama v. California: At the intersection of property, race, and citizenship. *Washington University Law Review, 87*, 979.

Wallerstein, N., & Duran, B. (2010). Community-based participatory research contributions to intervention research: The intersection of science and practice to improve health equity. *American Journal of Public Health, 100*(S1), S40–S46. https://doi.org/10.2105/AJPH.2009.184036

Wallerstein, N., & Duran, B. (2017). The theoretical, historical, and practice roots of CBPR. In N. Wallerstein, B. Duran, J. Oetzel, & M. Minkler (Eds.), *Community-based participatory research for health: Advancing social and health equity* (3rd ed.). Jossey-Bass.

Williams, D. R., & Collins, C. (2001). Racial residential segregation: A fundamental cause of racial disparities in health. *Public Health Reports, 116*(5), 404.

Wilson, S., Hutson, M., & Mujahid, M. (2008). How planning and zoning contribute to inequitable development, neighborhood health, and environmental injustice. *Environmental Justice, 1*(4), 211–216. https://doi.org/10.1089/env.2008.0506

Zuberi, T., & Bonilla-Silva, E. (2008). *White logic, white methods: Racism and methodology*. Rowman & Littlefield Publishers.

Index

© Springer Nature Switzerland AG 2022
R. J. Petteway, *Representation, Re-Presentation, and Resistance*, Global Perspectives on Health Geography, https://doi.org/10.1007/978-3-031-06141-7